HEAD, HEART AND HANDS IN HUMAN EVOLUTION

HEAD
HEART
& HANDS
IN HUMAN
EVOLUTION

R. R. MARETT

M.A., D.Sc. (Oxon)
Hon. LL.D. (St. Andrews)
Rector of Exeter College, Oxford
Fellow of the British Academy

New York
Henry Holt and Company

Printed in Great Britain

TO THE MEMORY OF
EDWARD BURNETT TYLOR

PREFACE

The title will, I trust, be found
to cover the somewhat miscellaneous contents of a book
which nevertheless has unity in the sense that its working
principles are throughout the same. After an Introduction
which calls attention to the rich variety of the fare awaiting
the student of human culture, Part I deals with the larger
issues of a theoretical kind involved in sociological inquiry
—in short, with the question how to use, and to keep, one's
head in the matter. In Part II we pass on to the central
topic of Religion, which, in its pre-theological phase at any
rate, is shown to be essentially an affair of the heart : Part
III being meant to afford particular illustrations of this
general contention. Finally, in Part IV, the material side
of human culture—in a word, the work of human hands—
is considered as it develops out of sundry rude beginnings.
Let me apologize to those of my brother-anthropologists who
are more especially concerned with technology if I seem to
be trespassing on their preserves ; but my attempt at a
bird's-eye view of the subject may serve a useful purpose in

7

the absence of any existing monograph that is at once detailed and comprehensive.

Part I embodies three Presidential Addresses, delivered to the Sociological Institute in the years 1933–35, two of which the Editors of the *Sociological Review* have kindly allowed me to reproduce with a few modifications. Part II represents the hitherto unpublished Donellan Lectures given by me in 1933 at Trinity College, Dublin, and afterwards repeated in a rather different form at Bangor. Of Part III the first chapter on Ritualism is a paper read at the International Congress of the Anthropological and Ethnological Sciences held in London in 1934 : it has subsequently appeared in *Folklore*, and I have to thank its Editor for leave to reprint. The second chapter on the Sacrament of Food was originally written for *Essays presented to C. G. Seligman* (Kegan Paul, Trench, Trubner & Co., 1934), and I am grateful to the Editors and Publishers for allowing it to see the light once more. The other chapters of Part III, together with the Introduction, are extracts from *Manners and Customs of Mankind*, edited by Sir J. A. Hammerton and issued by The Amalgamated Press, while Part IV is taken from another publication of the same Editor and Proprietors, namely, *Harmsworth's Universal History ;* and it is very good of them to let me put to a further use my occasional contributions to these two works of multiple authorship and encyclopædic range. In conclusion, if the reader find cause to complain that the treatment ranges all too freely between the academic and the popular—and, admittedly, the first half of the book deals mainly with theory, while the second half consists almost entirely of illustrative facts—let me venture to urge that the way to construct a true Science of Man can never be by depriving the subject of its human appeal.

R. R. M.

EXETER COLLEGE,
MAY 1ST, 1935.

CONTENTS

9

CONTENTS

INTRODUCTION. THE VARIETY OF HUMAN EXPERIENCE

Solitary confinement agrees with no man. Why, then, remain a prisoner when it is possible to range the wide world with a free intelligence and an open heart? Anthropology is the higher gossipry. It means literally " talking about people " ; and there can be no harm in such a practice so long as it simply bespeaks a friendly interest in one's neighbours. Over and above the sheer intellectual fun of surveying humanity at large, there is unlimited moral gain to be got in the enlarged consciousness of the fact that man is of one kind—that as a species we are near enough to each other in our type of mind to share all the thoughts and feelings most worth having. Moreover, whereas a good deal of our morality involves the disagreeable necessity of holding oneself in, the development of kindliness on a world-wide scale simply requires that we should let ourselves go. There is a latent power of sympathy in us all in virtue of our mating and herding instincts ; and to quicken it into full activity is the secret of human happiness, since in no other way can self-realization be achieved in so positive and ample a form.

There must, however, be method in our study of the diverse ways of mankind. Thus it is no use to trudge through endless galleries if one does not know how to look at pictures. To see a picture as a coloured board or canvas is not enough, since it must be apprehended and appreciated for what it truly is, namely a creation of mind, the outward expression of the artist's inward yearning for beauty. So, too, then, the mere surface-view of human manners and customs can yield nothing illuminating, and many a globe-trotting sightseer comes back no wiser than he went away. Strange customs must be interpreted in terms of their underlying motives ; though if this is done thoroughly most of the strangeness will be found to disappear in the process. Difficult as it may be to part with settled convictions about the general unreasonableness of foreign ways, a little reflection will make it plain that all human beings are striving, each after his own fashion, not only to live but to live well. Living, however, is an experimental affair, and of the numberless experiments that make up the history of our race a great many were bound to prove unsuccessful. Yet the worst failures leave no record ; so that we can reckon on having to do with at least partial successes, whichever way we look among known peoples, including even the most backward in culture. The humblest of them has hit upon a good thing or two, and so has managed to keep in the running.

Making due allowance, then, for the particular conditions, the judicious observer will always find something to admire —nay, more to admire than to blame—in any and every attempt of man to be master of his own fate. There is a magnificent hopefulness to be discerned in all the strivings of this importunate creature, who, as he emerges out of the mirk of the far past, is already scheming to work wonders— to confer immortality on his dead, to command fertility of nature, and so on. No facts are so hard that he does not seek to trick them into conformity with his visionary demands. Sole adventurer among the animals, he is ever ready to treat the unfamiliar as a manifestation of the divine, and is justified in so doing by the fresh access of vitality a richer experience brings in its train. It is his prerogative to convert mere cosmic process into a progress. He can affirm and will into being certain values which, though revealed gradually and always in response to much groping in the dark, are for him the most real thing in life—far more real than the passive conditions that must be overcome and transformed before his dream can come true.

This, then, is the only way in which to take an intelligent interest in human history—not to regard it as a cinematograph show, a moving pageant of diverting antics, but as the progressive triumph of spirit over matter. Slowly and painfully, but none the less surely, mankind has been fulfilling a creative mission in evolving out of a welter of blind instincts a " kingdom of ends "—an organized life-scheme centred on a supreme good comprising truth, beauty and righteousness, the three in one. No one can afford to be sceptical about this inner meaning attaching to the human struggle for real existence ; for, if he denies the good—denies, that is to say, our common interest in the making of it—he deserves to be eliminated. Luckily, however, there has always been plenty of good fighting material in the host of mankind ; and never is one more aware of this sterling quality of our race than when contemplating the first beginnings of culture. Every little savage group that keeps its flag flying in the face of the grimmest odds is a forlorn hope entitled to the greatest respect ; because at first all is confusion, and it is only by degrees that a battle array is established by forming up on some brave handful that, although unsupported, have managed to make a stand.

Hence, to do full justice to the anthropological outlook—in other words, to consider human history universally and in its true perspective—one must concentrate attention on that evolutionary inspiration—that divine urge for advancement —which lends its sole intrinsic significance to all our actions, and converts what would otherwise be an empty spectacle into an absorbing drama. At the same time, though this task of building up a moral universe is mightily serious at bottom, life has likewise its lighter side ; and, human nature being what it is, it almost looks as if too much earnestness might sometimes defeat its own purpose. Play serves as a preparation for work with the young of many an animal kind. Man has, however, immensely extended the sphere of his recreative activities until they intermingle with every interest of his adult life. There is a curious indirectness about the process whereby the higher culture has been in large part achieved. Many an unpractical pursuit yields solid advantage in the long run. In fact, all work and no play makes not only a dull boy but a stupid man. Human progress has all along owed more to the liberal than to the material arts ; in other words, the best way of looking after the body is to cultivate a soul. Science, fine art and even religion, with its symbolic procedure, transport the mind away from the plane of the actual, in the first instance ; so that when it has to come down to earth again it feels dissatisfied, and tries change after change in the hope of bringing the existent into closer resemblance with the ideal.

Thus it is not in the direction of the economic life that one must look for most of the extraordinary variety and colour displayed by human institutions as one surveys them in their vast range round about the world and up and down the ages. No doubt there have been devised in more or less direct relation to the food-quest a large number of appliances which are a credit to human intelligence, such as among hunting weapons the boomerang and the spear-thrower, the blow-pipe and the bow. Again, take fire-making—that decisive achievement of man which raised him once for all head and shoulders above the rest of living nature ; or the domestication of animals and plants ; or the smelting of metals ; or inventions so useful in their several ways as the canoe, the cooking-pot, the wheeled vehicle and the plough. Such clearly defined milestones along the highway of nascent civilization imply a tireless application of human ingenuity through long ages to the problem how to make the world a more comfortable place to live in.

Even here, however, where it seems a question of adapting means to ends of obvious utility, one wonders—and in default of definite information about these early victories over the environment one must continue to wonder—how much of the experimentation that led up to these fruitful results was initiated in a spirit of pure play. The poet goes too far when he suggests that " idle hands " can only work mischief ; for both manipulatively and mentally that fondness for fidgeting which we share with the monkeys can stand us in good stead, seeing that, unlike them, we have the good sense to follow up and perpetuate any novel experience that takes our fancy.

Man, then, in addition to utilitarian instincts that probably could do no more for him than keep him alive at the animal level, has an itch for the superfluous. Overbrained as he is, he has mental energy to spare, and can sport with trifles at no risk of vital ineffectiveness. Nay, since the liberation of pent-up energy is pleasant in itself, he is grateful to such relaxations in proportion as they afford him internal peace. Moreover, this discharge on to the play-object is not only soothing but at the same time stimulating, because of the self-revelation involved in bringing latent feelings out into the open. Now, when this happens in real life the agent is too busy with the affair in hand to dwell on the accompanying mental state. In play, however, which is but the mimicry of real life, the thing done hardly matters, and a corresponding mood can be appreciated for its own sake. The appetite for experience grows by thus savouring it consciously.

So we find the typical savage to be a mystic—the very opposite of the modern business man, who must have "real values " to deal with. An Australian greybeard will wander about the country for days carrying in his hand one of the sacred bull-roarers and " singing " the grass so that it may be persuaded to grow. No doubt he means it very seriously ; but in any case the operation is purely symbolic, and,

whatever its influence on the grass, has likewise the more
immediate and verifiable effect of making the operator feel
" full of virtue "—a spiritual condition so prized for its
uplifting quality that the older men, we are told, are wont
to make the acquisition of such experiences the chief interest
of their lives. Whereas, however, a trained intellect can
enjoy its inner resources in the form of more or less articulate
conceptions, the undeveloped mind of the savage must
perforce regale itself with crude emotions shot through with
but the vaguest thoughts and mostly excited and sustained
by vivid sense-impressions. To obtain this thrill of self-
awareness the primitive man has recourse to movements,
sounds and sights designed expressly to work him up to the
pitch required. The whirling dance, the throbbing drum,
the grotesque mask are so many stimulants which appeal
to him as much for their quantitative as for their qualitative
results. Indeed, his sentiment about all such means of
stirring up his faculties would probably agree with that of
the boatswain in *Treasure Island* as to choice of wine : " so
it's strong, and plenty of it, what's the odds ? "

It remains to note that this characteristically human habit
of clarifying and enlarging conscious experience by exercising it in leisure moments on toys and shams is largely
social in its origin. It is true that a negative way of inducing
ecstatic conditions favourable to the development of a certain
kind of self-centred personality is to withdraw from the
crowd and its distractions ; and of this or any other form
of voluntary abstinence it can at least be said that it helps
to give the will control over the passions. But such aloofness
is not congenial to the temperament of the ordinary man, and
least of all to that of the ordinary savage, in whom herd
feeling amounts to an obsession. His chosen method,
therefore, of intensifying his inner man is the positive one
of mutual excitation by means of what the Greeks knew as
the " chorus "—the collective ceremony or dance. The result
is, no doubt, to produce what is usually known as a group-
consciousness ; but it must not be forgotten that this exists
only in so far as it is individually shared. Because a large

part of our education has been acquired in class, it does not follow that we have no right to claim it as our very own. Indeed, all culture is traditional, that is, a social inheritance, as it first comes to us, though we may hope to pass it on to others a little different and better for having been transmitted by way of our minds. Hence a savage cannot be denied to have a private soul because fellow-feeling forms the core of it. Nay, just because this is so, the civilized observer is apt to regard primitive ceremonies as devoid of meaning, not realizing that so long as they engender the sense of fellowship little more is asked of them by the participators, who, after all, are principally concerned in the matter.

When it is a question of a more or less definitely religious rite of the primitive pattern, we should be wrong in assuming any consistent doctrine to underlie the performance. The general nature of the blessing sought may be indicated to some extent by the occasion, as when rain is wanted, or a disease has to be banished, or a marriage must be solemnised, or the dead man laid to rest. But the rich medley of forms that has come down from the past as appropriate to each recurrent situation is, as it were, repeated by rote and without any attempt to make sense of the separate elements of the mystery. It is a common fallacy to suppose that the savage has forgotten what it would be truer to say that he never tried to understand. A play of images sufficiently forcible to arouse by diffused suggestion a conviction that the tribal luck is taking a turn in the required direction is the sum of his theology ; and yet the fact remains that a symbolism so gross and mixed can help the primitive man to feel more confident of himself—to enjoy the inward assurance that he is in touch with sources and powers of grace that can make him rise superior to the circumstances and chances of this mortal life.

As regards the persistency with which the human mind clings to any well-tried device for stirring the emotions through the imagination, much is to be learnt from folk-lore, a branch of anthropology which has sometimes been defined as the study of survivals. Now it is true that the folk, by which is meant the relatively uncultured and, in particular, unlettered portion of a civilized community, is itself on the way to extinction ; so that its customs are necessarily taking the same downward path. Nevertheless, along most of the European country-side and even in parts of America—among the French Canadians, for instance, or the mountain-people of Kentucky—there are remains of an old-world tradition propagated by imitation and word of mouth on which, however secretly, the majority depend for a good deal that gives their life its flavour. Some of it, indeed, can have no other function than to keep alive feelings unworthy of an enlightened society—the fear of witchcraft, for example, and kindred superstitions. Many folk customs, on the other hand, are and have always been quite healthy in their moral tone, and in essence are diversions, providing, as it were, relief from the common round with its soul-destroying because all-too-absorbing insistence on the needs of physical existence. Whereas a misguided puritanism once dealt hardly with maypoles and the like, it might as well have declared that Milton's poetry was pernicious fooling, because this too sought a wider horizon by mounting on the wings of fancy.

The same kill-joy attitude towards many a perfectly harmless institution of the savage is to be noticed to-day among ourselves, who profess to be engaged in improving his condition and, on the whole, may be taken to mean what we say. Clearly, if he loses more spiritually than he gains materially, we shall have failed in our mission. Our only chance, however, of making a new and better man of him is to leave his highest values intact, so that by preserving this foundation we can build solidly upward bit by bit. Our first duty as educators, then, is to leave him a play-world of his own. It is certain that in the past he has found himself chiefly in and through his ceremonial life. Let not his future be rendered as drab as that of many a drudge of civilization by cutting him off his spiritual holidays, which he has a right to spend just how he himself chooses.

Let it be agreed, then, that spiritually it takes all sorts, as the saying is, to make a world. Difference is not incompatible with unity, but, on the contrary, a diversity of elements yields the richer harmony. On the economic plane we are gradually moving towards a relative uniformity of conditions. The time has gone by when each people had to make what it could of the material resources nearest to hand. Widened communications have facilitated the distribution of the means of life until there prevail common standards of comfort such as are bound to provoke a no less universal demand for their satisfaction. Even the differentiation of habits imposed by climate has been largely counteracted in the modern world by ingenious protection of the body ; so that, in its physical aspect, human life throughout the globe is threatened with a certain stupefying sameness. All the more reason, then, is there to give free scope to our natural variability on the side of the spirit.

In this direction there are always discoveries to be made, new worlds to conquer. Whether in science, or in art, or in religion, which at its best should comprise all the higher interests of life, the mind is not merely exploring experience but is literally creating it. Here lies the true work of man : having secured his modest place in the sun, to devote all the rest of his superabundant energy to accommodating himself within that region of inner light where he can live so much more spaciously.

The upshot of such considerations would seem to be that we should be tolerant of variety in human affairs and suffer strangers gladly. Behaviour, however odd it may seem, must be judged by its intention. First impressions of the meaning of a kindly act are often misleading, as when by way of salutation a savage sheds tears, or offers to rub noses, or actually spits in one's face. Given tact and a sense of humour it is always possible to adapt oneself to the local code of correct manners ; and it is a mark of inhumanity to refuse to do so. Or, again, it is sheer impertinence to want to interfere with customs in dress, even if some primitive fashion-plates subordinate concealment to emphasis, as, indeed, do some of our own. Indeed, as regards all matters of taste, one can heartily endorse Kipling's assertion that

> There are nine and sixty ways of constructing tribal lays,
> And every single one of them is right.

They are right, that is to say, for others, if not for us—essays not to be denied the right of publication because they may or may not find many readers. Any number of vehicles of expression will serve to convey the same message, and it is indifferent which of them is used so long as the meaning comes through with sufficient clearness. Just as one can learn to think in more than one language, and by so doing becomes all the more competent to distinguish sense from sound, so acquaintance with different cultural conventions helps the mind to detach essential values from their casual context and to measure them by the sole test of their intrinsic worth.

At this point, however, one will be reminded that there is an evolutionary process—that even of ideals it is true that only the fittest survive. Granting this, one may at the same time insist that the evolutionary principle be interpreted correctly. Thus, the so-called " unilinear " version of it is quite out of date. We human beings are not advancing in one long line up a beaten track. Rather we are thrown out like skirmishers along a wide front, making for an unspecified objective, but obeying a general order to go forward. In these circumstances it is hard to say which part of the line is ahead of the rest, since rapid movement will not make up for loss of direction. In practice we look back over our shoulders to count the dead, and decide that the least dangerous track will lead us farthest. Meanwhile, it cannot be said that it is reserved for the big battalions to make real progress, since imposing civilizations in the past have been overwhelmed in sudden and wholesale disaster.

To assume, then, that we are culturally and even racially the dominants destined to lead the species to final victory is decidedly premature. Certain it is that mechanical aids will not compensate in the long run for lack of morale, which perhaps is not our most conspicuous asset. Have we solved the marriage problem? Do we know how to educate the young? Can we produce intelligent government, or a noble art? Does our religion concentrate on essentials? That we can propound such questions to ourselves in a spirit of sincere self-criticism is perhaps the most hopeful sign of the times. But such recognition of the need of putting our own house in order carries with it an obligation to be chary in passing moral judgment on our neighbours, even when from our position in the line they seem to be hanging back. We have seen reason to concede them a right to their own manners. How far, then, can we extend the same privilege to their morals?

Hard saying though it may be, it looks as if it might be better for all, including ourselves, to treat ethics as an experimental science and, within limits necessary for the common safety, to allow and even encourage a diversity of moral institutions. Indeed, on grounds of policy only the British Empire has to treat on a more or less equal footing distinct and even rival religions, systems of law, and so on. While policy approves, need conscience protest? Surely, if it be allowed that morality, or religion, or law, or any other scheme of values evolves or has a history at all, then it follows that alternative possibilities should be thrashed out if the selective process is to be thorough. As we are prepared to advise our neighbours, so let us be ready to take a hint from them on occasions. Even the so-called savage—a word which originally had the innocent meaning of "woodlander" —is not a mere Caliban, but can sometimes provide so edifying an example of the simple life that sympathetic observers such as Alfred Russel Wallace are set wondering whether it is possible to be both civilized and good. Even when primitive ways appear at first sight less edifying than shocking, as, for example, when we come across customs of courtship and marriage highly offensive to our taste, it is well to look carefully into the facts, when we may well be led to view the matter in a new light. Thus, we are horrified to hear of a savage buying his wife. When it turns out, however, that by means of what is misnamed the bride-price he is simply purchasing the right to enrol the woman's children in his own clan, the practice seems reasonable enough ; and one is perhaps led on to take to heart the old-world view that the end of marriage is to have a family.

So much, then, by way of introduction to a series of studies that show our race engaged in some of their more striking and characteristic attempts to transcend the level of brute existence—to snatch a grace from their life by investing it with meaning as the pursuit of a spiritual good. To realize how much steadfast trial and bewildering error went to the discoveries embodied in this diversity of custom is beyond the reach of any imagination. Yet everyone can at least study the human record with a proud sense of the pluck and determination that have carried mankind along from strength to strength, eternally dissatisfied, eternally hopeful.

PART I.
THE SOCIOLOGICAL OUTLOOK

1. EVOLUTION AND PROGRESS

A man of science ought to know how to mind his own business. If he chooses to play the philosopher in a leisure hour, as he has no less a right to do than any other man, he should at least make it perfectly plain to himself whenever he is straying beyond the limits of his official task. This caution applies with equal force to his use of categories or supreme heads of classification. If they exactly coincide with the full extent of that particular subject, then well and good ; they are clearly his, to be employed as he pleases. If, however, he is importing wider generalizations into the investigation than are strictly necessary for its independent conduct, he is inevitably begging certain questions of the philosophic order. Whereas his constitutive categories are his very own, these wider categories, which may be called regulative, are borrowed. It is in the capacity of a more or less amateur philosopher that he has recourse to their help. Even if he adopts them uncriticized and ready-made, as probably happens, he must shoulder the responsibility none the less if they turn out to be misleading rather than helpful within his proper field of research. The least that he can do, then, is to use the several kinds of principles with discrimination. He must be able to distinguish clearly between the string that ties up his private bundle of facts and the stouter rope needed to fasten it together in a bale with a number of other such parcels.

It may be worth while, then, in what follows to consider from the sociological point of view two categories that are rather easily confused, namely Evolution and Progress ; the former being of the regulative and the latter of the constitutive type. Progress is an idea applying to men only ; whereas Evolution applies to men and to fleas, and, some would say, to stones as well. How far, if at all, mankind is progressive is a question which it is entirely within the province of the sociologist to answer. On the other hand, whether mankind is evolving can only be settled in committee with the representatives of other sciences, who would do well to have a trained philosopher for chairman. To consider the wider issue first, why does the sociologist normally proclaim himself an adherent of the evolutionary principle ? There are at least good historical grounds for this attitude. Sociology owes, if not its very existence, at any rate the secret of its scientific inspiration to its association with Darwinism ; and Darwin called himself an evolutionist. Wisely or not, he was prepared to avail himself of the category of Evolution, although for him it was regulative rather than constitutive in function—in other words, covered more than his special purpose. For it was Herbert Spencer, not Darwin, who invented the term, having a great deal more than biology in his eye when he did so. Spencer was, in fact, trying to formulate a law embracing mind and matter alike as twin constituents of the universe. Thus as a biologist Darwin might well have hesitated to borrow another man's phrase from such a philosophic context. Biologically speaking, there was no need to postulate more than was strictly required to explain life as such by summing up its processes under one supreme formula. Literary convenience is hardly an excuse for overstepping the limits of a purely biological argument in this way. Whether increased complexity of structure in an animal or plant is on wider grounds comparable with the change that occurs when condensation in a nebula gives rise to a star, is a consideration wholly beside the point when the origin of species is being discussed on its own merits. There is all the difference in the world between Spencer's Evolution as used technically with a cosmic reference and Darwin's Evolution in so far as it stands loosely for a way in which living things behave. Whatever personal reasons Darwin may have had for admiring the Spencerian philosophy, it is very doubtful if he seriously meant his biological theory to depend on this framework. In view of his characteristic modesty one cannot suppose that the father of modern biology claimed to be more than a man of science working in full accordance with such methods as are strictly appropriate thereto.

The fact remains, however, that by borrowing the term Evolution from Spencer, Darwinism seems to be fraternising with physics or, it would perhaps be more accurate to say, with a physical brand of metaphysics. In science all such borrowing of categories stands for a friendly act as between allies. By calling itself "evolutionary," biology acknowledges affinities that might well cause it to be regarded as a kind of bio-physics. It may indeed be that certain biologists of the behaviouristic or anti-vitalistic school would frankly have it so. Living matter, they would suppose, is just matter that behaves in a peculiar way. Now they have a perfect right to make this guess so long as they do not forget that the peculiarity in question constitutes the only aspect of the case on which they themselves are entitled to give an expert opinion. The properties of matter as such are not, in the last resort, their affair. Hence, if they whittle down organic process until nothing but physical process is left, their subordinate services are no longer required. The patient having been duly persuaded to succumb, the doctor's job is over as undertaker's jackal. Thus the formula that evolution is change from simple to complex, if supposed to sum up the life-process, can represent it merely as a certain complication of matter subject to further complications on its own account. The remarkable fact that it has to be credited with an account of its own has to go by the board in order that this intrusion of a physical category shall pass unobserved. To grow more organised cannot mean, off hand and on the face of it, the same thing as to grow more complicated. However mysterious and awkward it may be for the investigator, there is a factor in the problem which happens to be the distinctively biological factor : and what is wanted is a formula comprehensive of that. Darwin may have simply meant development when he spoke of evolution ; but, if so, he unfortunately used a borrowed phrase, a metaphor, such as suggests an unholy alliance between biology and physics, with downright materialism somewhere in the offing.

At the same time there is no need to treat matter as a bogey. Such a conception is quite useful in its place. Matter is in fact the democratic category *par excellence*, because it appeals to the commonest, if not necessarily the best, kind of experience that human beings enjoy. The surface-view of things afforded by the senses, and more especially by those of touch and sight, commands a fuller publicity than any other. What usually goes by the name of common sense consists in just this tissue of communicable externalities. No doubt the physical and quasi-physical sciences do much to reshape it. But they have no power to alter the essential character of the stuff on which they work. Sense-experience remains the current coin of human intercourse, whether it bears the official stamp of science, or consists of baser metal that nevertheless manages to pass with innocent folk. The practical man, then, will look on matter exactly as he looks on money—namely, as a convenience for the wise, however easily turned by fools into a fetish. By compounding our sense-impressions, and attending simply to what may be called the mathematical qualities that they yield for the construction of objects, we can create a system of cash-values whereby all the rest of our experiences can, however roughly, be compared and equated for purposes of universal exchange. What is known to the natural sciences as verification, or making a thing true, is the result of bringing that thing as directly as possible within the range of touch and sight, so that it can thereupon be established firmly within the daily experience of the multitude.

If, then, there is an aristocratic type of soul with experiences of its own not so readily shared, so much the worse for average humanity, which just in so far as it is sense-bound will tend to regard matter as the only touchstone of truth. Thus catholicity is a pretension of religion which in practice it by no means succeeds in making good ; whereas science of the physical or naturalistic type has everything to expect and nothing to fear from vulgarisation. The beliefs precipitated out of the spiritual strivings of the genius in religion, philosophy, or art are not often imparted without distortion even to the most intelligent of his disciples. On the other hand natural science, if it fails to make its meaning clear to all, has only itself to blame for using a verbal shorthand ; or else for slipping unawares into metaphysics, and admitting non-sensuous symbols at the point where actual sense-experience breaks off and exhibits a ragged edge. There are doubtless excellent historical reasons why a materialistic outlook should prevail. The most refined of the human faculties are relatively late-born, whereas the routine of the senses was organised a long way back in time. Our first ancestor started his career with a sense-apparatus which in most respects, even if smell was a weak point, must have been on a par with that of any of the higher mammals ; while he had the extra advantage of being able to help out his eye by means of a very mobile hand. Common sense and handiness between them are thus the earliest assets of our race. Man could specialise from the first on matter and machines. By bringing our own bodies, mainly by way of the hand, to bear on other bodies, we have discovered how to modify their working to our advantage in all manner of ways ; and that without treating them as anything more than bodies. What is known as our material culture consists entirely in the sum-total of these exterior manipulations ; and it is to be noted that it is such material culture which is most eagerly adopted, and hence most easily diffused, as between different portions of mankind. In this communicability, then, alike for purposes of knowledge and of use, is a claim for recognition on the part of the material aspect of the universe which it would be very short-sighted to belittle.

Those, then, who would champion the spiritualities in face of this world-wide and age-long appeal of matter must be careful not to set about it in the wrong way. If they try to kick matter out of their path they will simply stub their toes. Matter is a solid reality, at any rate in the pragmatic sense that it has the solid consent of the crowd behind it. If votes are to count, there is not the slightest chance of obtaining a resolution to the effect that the whole realm of sensible experience is one vast illusion. A counter-motion maintaining that everything without a sense-backing was empty dream-stuff would be much more likely to be carried by general acclaim.

Indeed, if such a false issue were to be raised, it is hardly doubtful on which side the psychologists themselves, who ought to be the professed defenders of the soul and of the things of the soul, might be inclined to enlist ; for many of them would wholeheartedly make common cause with the public. After all, it is a very crude psychology that treats soul or mind as something completely alien to body. To discard matter one must, in order to be consistent, like-wise discard the body, together with the senses so far as they depend on the body. If after these rather impracticable eliminations were accomplished aught were to remain, it would certainly be an ineffable kind of experience, since all possibility of speech would have been thrown overboard with the rest. In fact, there is but one *experimentum crucis* possible in the circumstances, namely, dying ; and even *revenants* show a suspicious predilection for the use of a corporeal medium. It is quite another matter, however, to agree, as the modern psychologist would do, that there are degrees of what he would probably label introversion to which the average sensual man seems incapable of attaining. A sort of spiritual levitation marks the genius, whereas the majority must cleave to *terra firma*. Thus, though the realism of the groundlings may well suffice for them, there is room also within the mind of the more enlightened for a high-grade consciousness, which, without rejecting the realistic point of view, nevertheless transcends it ; that which is superadded being essentially an idealistic, reflective, inward-seeking power. As sight is to insight, so are the materially-minded to the spiritually-minded, the many to the few.

How, then, is biology going to do justice to the psychological element which life includes, more especially when this element seems inclined to take charge of body and to use it for its own purposes rather than to conform passively to the ways of body as such ? The biologist who is not likewise a sociologist deserves all our sympathies, because with the best will in the world he cannot make much use of psychological aids in the study of our dumb companions, whether animals, insects, plants, or something between the three. He is working backwards from human analogies when he attributes consciousness, intelligence, and so forth to lower and ever lower organisms. These nevertheless one and all exhibit a certain purposiveness in their bodily reactions to other bodies, live or unalive. Hence he is tempted to treat life as coextensive with mind, though in some minimal sense of the term life need connote no more than pure instinct. At this point he hardly knows which way to turn. Shall it be towards mind ? But mental activity becomes ambiguous in proportion as it departs from the type of which we have introspective evidence in ourselves. As we read back consciousness through the hardly conscious into the quite unconscious we are at last faced with what Lewis Carroll would call " the grin without the cat." Shall he then turn towards matter ? But organism is something more than mechanism. In appearance it is a self-determining unity rather then merely the joint effect of certain convergent forces. No system of external compulsions is likely to prove

31

an attractive assumption when it is so much simpler to suppose that every creature with a tail is itself responsible for wagging it. At the same time, even when it takes the form of bare instinct, life cannot be set apart by itself as something neither mental nor material but simply neutral. It cannot be neither, so it probably is both. Body is obviously involved, and mind, though less obviously, cannot but be involved too. We are in fact at a half-way house : yet, once there, surely the impetus of the search for a higher principle will lead us to go forward rather than to go back.

For, alone among biologists, the student of the life of Man has the chance of making full use of mind as a principle of explanation at least as valuable as body. Whatever may be the case with his less fortunate brethren, he has access to a master-clue. No fox without a tail is going to persuade him to dispense with an appendage at once so ornamental and so useful. It would be absurd to insist on levelling down the man to the bacillus for the simple reason that the introspective instrument is lacking for levelling up the bacillus to the man. We must surely be ready to embrace with due gratitude the single chance presented in our own case of getting an inside view of reality, as at any rate represented by that part of reality which we in our own persons not merely embody but likewise ensoul. Primarily the sociologist needs mind or soul as a constitutive category to be used for working purposes as at least on a par with body, whatever predominance the latter principle may enjoy in less favoured departments of biological science. As for using mind as a regulative category based on some estimate of the part played by it in experience as a whole, this would be definitely to shift over to the higher and more perilous ground of philosophy ; and such a step must not be taken unawares. It is, however, plain that the natural affinities of sociology, with its unique command over the psychological factor in the life-process, will be with such a metaphysic as attaches due importance to the mental side of reality. The inferior court of sociology is competent to try the case of human soul *versus* human body. Its decision, however, which is likely to be something in the way of a compromise, can count at most as but a sort of precedent in that supreme court which, in the rough and ready style of the law, would settle the more far-reaching dispute between mind and matter. As for the possibility of obtaining an impartial verdict—that is, literally something *veredictum*, or said with the finality that belongs to truth alone —it cannot be denied that mind, and, in some sense, the mind of Man, must act as judge in its own cause in both inquiries alike. The choice, however, would seem to lie between a living judge and a mere lay figure, some dumb idol counterfeiting truth and justice, and man-made at that. If matter wins, mind will have endowed its own figment with the power of silencing all opposition by its grand incapacity for saying anything at all.

To address ourselves, then, to the simpler question whether any progress in respect to body and mind alike can be made out on a broad view of human history, there seems little doubt that the answer will be in the affirmative, if we treat body throughout as the subordinate factor in the problem. If we make what the biologist knows as dominance the test of progress, without for the moment considering the inward condition of a richer experience that goes with it, who would venture to deny that, compared with his far-off or even his recent predecessors, the modern man has increased his numbers, and therewith his grip on the surface of this planet ? Continuous observation from Mars or Venus assisted by reasonably powerful telescopes could not have failed to observe modifications of the whole terrestrial bioplasm for which Man and the weather between them can take most of the credit, and perhaps Man even chiefly. Now this gain in effectiveness is not very visibly connected with change of body as such, though it must be remembered that we have great difficulty in observing the finer alterations undergone by the brain, the directive centre of our energies. Indeed, there are some who would, on grounds of pure physique, contrast the man of to-day more or less unfavourably with earlier types of *Homo sapiens*, such as the stalwart Cro-Magnon ; while even the latter might have been

C

worsted in a wrestling match with that Neanderthal man whom nevertheless he managed to outwit or at any rate to outlive. In sheer muscle, then, we may have actually degenerated, even while securing in other ways a handsome balance of advantage. Nay, it is certain that we abuse this advantage by gratuitously permitting non-selective breeding to fill our ranks with weaklings, who cannot but impede the onward march of the race.

Yet this lamentable neglect of eugenics may be taken as affording a negative proof of the fact that the chief source of our strength lies in the soul rather than in the body. We tolerate dysgenic customs because the development of social sympathy has hitherto proved incompatible with public regulation of the individual right to marry, and to have children regardless of their physical or even their mental quality. The undoubted drawback of supporting a crowd of defectives on a permanent dole can be ignored only on the excuse that to curtail our charity would be the graver risk. Certain it is that the truest measure of human progress as revealed in history is the growth of culture in what may be broadly termed its moral aspect. A superficial reading of the human record will doubtless tend to emphasize the materialistic aspect. The more important technological inventions, by bringing about far-reaching changes in the economic life, have obviously acted as powerful levers in their way. But it should not be overlooked that every society can only acquire the degree of civilization that it deserves. Savages have been known to wash down a cannibal feast with draughts of trade-champagne ; and so too there are more aspiring peoples who cannot altogether match their manners with the style of the goods which they are able to buy or to copy. The present age has in fact seen a vast diffusion of material improvements, unaccompanied by any propagation of higher moral principles on a corresponding scale. The tide of human progress, however, is subject to alternating phases of reflux and advance ; and it may be that after a great war, which was essentially a struggle for markets, the exhausted nations have taken to heart the lesson that competition is not always beneficence in disguise, and will bring themselves to try whether good faith and good will do not offer firmer foundations for a lasting civilization. This is a crisis in human history when our educators and our statesmen—those of them at least who are lofty-minded, and have the courage to make public opinion rather than to follow it—must preach the gospel of universal amity, with that sublime contempt for the practicable which is always the mark of inspiration in ethics and religion alike. "Though we cannot, we must" is the paradoxical declaration of will that pre-conditions all high adventure. History shows that the merely existent will always submit to transformation under the magic spell of the ideal. 320205

Common sense, then, is by no means the saving virtue of the true leader of society. The crowd after all has it in plenty : and its inveterate materialism acts as a drag on moral progress, which in any case is collar work for all concerned. The seer, for the very reason that he denies the actualities of current experience and offers visionary possibilities in their place, must always be by common standards something of a madman. Judging by those same standards one might even say that, if he is wholly sane, he must be a charlatan. Just as no genuine artist can create to the order of the philistine, so in ethics and religion the pioneer who has " hitched his waggon to a star " is incapable of reining in to suit the pace of the average laggard. Fortunately, the impatience with genius ever manifested by the less enlightened is normally accompanied by a certain awe. So long, then, as this feeling happens to be uppermost, a contagious faith in the personality of the prophet will be apt to gain a like credence for his message. It is common to all the hero types which Carlyle tried to distinguish that they should alike be deemed worthy of worship by the multitude of their adherents. One and all they are felt and believed to have *mana*. They are acknowledged wonder-workers— not manipulators of matter, but dealers in the unseen, men who summon latent things into manifest being by the power of a master-word. The source of this dynamism of theirs

35

remains for most men a mystery, not so much because the possessors of the secret would guard it for themselves—since on the contrary they would have their inward light be likewise a light for the whole world—but rather because they openly bear the stigmata of a painful initiation, such as is to be avoided by those who prize their creature comforts. Among the host of lesser men who fill the pages of so-called history with their clatter it is often hard to distinguish the authentic heralds of a new moral order, apart at all events from a few outstanding founders of religions or quasi-religious philosophies. Nevertheless, in any commemoration of the benefactors of our race such rare spirits deserve chief mention; and any account of human progress, not cast on vulgar lines and blatantly satisfied with triumphs over matter yielding ponderable results, must lay stress rather on the imponderable increment derived from moral and religious education, whereby human life has gained so much more in the way of deepened experience, and hence of positive reality.

That progress in the direction of the spiritual is implicit in normal human endeavour would seem to be the moral of history, even when carried back to Glacial times. Man was always ready to turn aside, as it were, from the business of mere living in order to cultivate methods of living well; and counted it no loss of time, since he thereby found himself. For example, the fine art of the Upper Palæolithic cave-man belongs, as no fine art can altogether cease to do, to the level of sense-experience; and its naturalistic character indeed reveals the closest association with the keen eye of the hunter in search of a meal. Nevertheless, beauty is something perceptible only to the eye of the soul—to a faculty of inward vision, which clothes the outward object with an *aura* invisible to the man who can only see his reindeer or bison as prospective meat. The mind of the artist is already at one remove from that surface-view of things which calls itself realism and is actually a shallow phenomenalism. So too at a later stage Babylonian and Egyptian star-lore, despite its astrological tendencies, must have generated a disinterested interest in the movements of the heavens;

though it was left to Greek science to glorify this as an ecstasy of pure " theory," that is, contemplation. Thus an intelligence free to pursue its own ends for their own sake was gradually liberated from its enslavement to the workshop. Nay, the previous astrology with its magico-religious purpose was always divided in its allegiance as between the practical and the transcendental ; and, in regard to sciences of the physical type in general, it might plausibly be argued that they originate in moral attitudes towards super-physical agencies, and acquire their reference to the sense-world mostly later and by the way. For Man, however primitive he may be, sets such a high value on his moral relations as developed within the home-circle, however narrow, of family, kin and tribe that he all too naïvely expects a like sympathy from a personified Nature ; and makes overtures of friendship to the universe at large on what might be termed a pan-ethical basis. Alas, sad experience in the long run convinces him that, whereas some of the hard facts of existence—the men of the next tribe, for instance—do eventually respond to such treatment, others do not. Indeed, the problem whether the weather may be affected by prayer may even now be regarded as in the experimental stage. A cocksure materialism is, in fact, a disease of thought that has attained to epidemic proportions only in quite recent times ; though religion is partly to blame for having driven common sense into opposition by seeking for the spiritual in the wrong direction, namely, without instead of within.

Taking the auspices, then, with the help of the long-range telescope of history, one can discern a future for humanity consistent with a progress in spirituality along lines for which its natural impulses and aptitudes have designed it from the first. Such progress may be conceived in terms of the greatest self-realization of the greatest number. Not happiness, which is too closely bound up with sensual pleasure, but rather a kind of blessedness, an inbreathing of a diviner air, is the only true object of moral education, which in turn can be identified with progress, so far as such a striving after perfection can be treated as means and end in one. For the reward consists in the effort itself, when it is an effort to turn becoming into being—to identify the self with certain timeless values which it can desire, know, and will, without impediment to its continuous activity. We can enter well enough into the experience of the saints and sages who testify to the reality of these abiding satisfactions to be sure that human nature can touch these heights. The trouble is to discover how far these same heights are accessible to less hardy and resolute pilgrims.

Now, whatever one's bias in favour of political equality, souls are not to be confounded with statistical items. At most it might be urged that all are equally capable of further self-cultivation, though by no means after the same manner or to a like degree. Thus, while sense-capacity offers the best example of a racial constant, no known tests will bring out a standard reaction to beauty, even when reinforced by the universal appeal of sex. Hence it may be doubted how far the most intensive education in the fine arts can effectively popularise what might be likened to a veritable gift of second sight, wherein whole nations, at any rate during long periods of their history, are liable to be defective. Again, intellectual ability varies considerably as between individuals, not to introduce debatable questions concerning race or sex ; and not every climber can leap the gulf that separates a pass from a class in philosophy or in the higher mathematics. It must be added, however, that there are so many fields of inquiry within which thinking, both analytic and synthetic, can find free and noble exercise that in this direction there may be greater scope for interesting the wider public ; the more so because team-work is here far more possible than where fine art is concerned. There remains, meanwhile, a third and greatest sphere of education, from which none but the morally infirm are constitutionally debarred. For all the other refinements of the soul can be brought to bear on the dissemination of love among mankind. The seed exists, but has hitherto been grown in small patches ; and it will need a scientific agriculture, of which the very rudiments are barely understood, if a love-starved world is to have its fill of this life-sustaining food.

Adopting, then, a slightly altered version of the Platonic demand for the philosopher-king, let us say that moral progress will be not only possible but inevitable as soon as our statesmen become educators and our educators statesmen. Long ago Mill based his defence of representative government on the ground that its educative value outweighed any failure on the administrative side. The same should surely hold of every kind of government worthy of the name ; and its educational policy should be ever to the fore in the deliberations of the ruling body. Nor is it enough that a nation should be indoctrinated with the science of political economy, which has always tended to ignore its duty to subserve the supreme art of politics, preferring to model its methods on those of the naturalistic or unethical sciences. Everything should be done to encourage all higher studies, and all fine arts ; while the planning of noble monuments, and the preservation of natural amenities, should be made a further means of elevating taste and promoting travel and intercourse. By such peaceful paths a spiritual advancement, individually initiated it may be, but even so capable of being collectively shared, can be confidently predicted for mankind ; so that in the end war will be rated at its true value, namely as a means of finding work for the morally unemployed. As for religion, which claims by divine right to preside over these higher activities and to crown them with a super-humanitarian sanction, it must put its house, or rather its many houses, in order. If it would effectively govern instead of nominally reign, it must seek in a broader charity the remedy for its long-standing and almost constitutional malady, which consists in sectarian exclusiveness conjoined with a mania for persecution. No marriage between church and state is likely to have happy results so long as the better half remains a mixture of snob and shrew. Nevertheless, there is nothing that offends either moral aspiration or historical probablity in the prospect of a future world religiously governed, in which the gentler

elements will prevail, sublimating the crude energy of the
fighting male into forms of virtue more directly subservient
to the upbringing of the human family. For, if the present
argument be at all sound, here are the facts. Real progress
is progress in charity, all other advance being secondary
thereto. The true function of religion, on the other hand,
is to enlarge charity, and therein is summed up its duty both
to God and to man. Because the religions of mankind, like
their languages, remain irreconcilably plural, the sciences
and arts, being more catholic in their appeal, have at present
taken charge of those spiritual interests that alone can knit
the nations together. No central organ exists to bring these
nobler tendencies to a head. Yet neither politics nor ethics
can supply this, apart from a common religious faith that,
in helping ourselves to become more perfect, we help the
ultimate perfection itself to be justified.

Having done our best to find a reasonable use for the category of Progress as a key to the movement of history, it
remains to ask ourselves whether Evolution, in either its
Darwinian or its wider Spencerian connotation, can invest
this purely human tendency with a more general applicability,
and hence a better claim to explanatory value. The position
would seem to be this : that we are in possession of a good
thing, known at first hand and inwardly by Man, the
supreme judge, to be good not only on the surface but all
through ; and that we are not going to give it up because
apparently the rest of the universe, organic or inorganic,
either has got less of it or has not got it at all. Progress, in
short, cannot be declared null and void merely because it
pertains to spirit, and spirit is forsooth an aristocrat, whereas
matter is democratically all on one dead-level. Remove
spirit from the universe, and you remove all meaning whatever, even the meaning attaching to the difference between
simple and complex, or to any other version of the evolutionary formula. On the other hand, if spirit is to provide
the sole possible criterion consisting in the triple test of
intellectual meaning, moral value, and metaphysical reality,
it can only do so out of the fullness of our experience which

ultimately classifies all things in terms of their relative goodness ; the most comprehensive kind of goodness, in other words the richest in meaning, value and reality taken together, being self-evidently the best.

A philosophical interpretation of first principles, then, will allow each science to determine its constitutive categories as it pleases, but will reserve the right to arrange these categories, and consequently the sciences that use them, in a hierarchical order according to their relative comprehensiveness. Thus any one of them is regulative only in regard to some lower member of the ascending series. It turns out, therefore, that it is not Evolution that is regulative in relation to Progress, but on the contrary that Progress, being Evolution and something more, has the right to impose on Evolution such interpretative limitations as may be philosophically necessary. It can say, for instance, to the biologist who tries to explain life-process in terms of natural selection that, as his very choice of analogy suggests, conscious selection, which is a constitutive principle of sociology, is a fact of fuller significance ; and hence that natural selection is at best but an unconscious kind of selection, so far indeed as we dare read into it any true selectiveness at all. So, too, again, if at this point the physicist begs to take over the burden of explaining a remainder that grows thinner and thinner as difficulties are cleared out of the way by the convenient device known as abstraction, philosophic criticism is obliged to insist that matter is, regulatively speaking, a sheer negation of life ; though constitutively regarded it may suffice to provide a groundwork for mapping out certain residual relations of the time-and-space order.

In short, Evolution in whatever sense it is used is, from the sociological point of view, a useful servant but a bad master, being the inferior category as compared with Progress, that distinctively human attribute which history proves to be a real thing, when its spiritual nature is duly taken into account. As a category of ordinary biology, Evolution is subordinate, because by its inclusiveness it relates to things that are largely below the human level. Hence it cannot be regulative in a sense that would enable it to rule out what occurs at the human level, as if that were only a superfluous demonstration on the part of the universe, and one which its best friends think it only proper to ignore. If, on the other hand, the idea of Evolution is extended so as to cover Man's continuous gain in what his intelligence assures him to be spiritual worth, then we must read backwards into the beginnings of life, and perhaps even into the prior labourings of so-called lifeless matter, a certain gradual awakening of a force instinct with those magnificent possibilities that are already within the reach, if hardly the grasp, of Man, the most godlike of earth's creatures. Evolution must in that case be conceived as the liberation of spirit, by means of its own activity, from a sort of sleep spent at first almost dreamlessly within the womb of time, and even now yielding but slowly to the efforts of the mind to assemble its faculties and be itself. Such a view of life, instead of treating it as an accident of a time-process itself unsubstantial since without meaning of its own, makes it the expression, through the mechanism of the body, of an immanent intelligence and will ; of whose ultimate purpose we as social beings can at least judge well enough to say that we are most in sympathy with it when we are most in sympathy with one another.

2. FACT AND VALUE

In seeking how to keep one's head in the study of Man, I make no apology for again calling attention to first principles. We sociologists are notoriously vague, not to say confused, in respect to our architectonic. For this indefiniteness of outlook, however, we can offer plausible excuse in the sheer size and complexity of our subject-matter. This being nothing less than the social factor in human life, we find ourselves committed to the comprehensive study of all that is covered by the term " culture "—or " civilization," as we call it in so far as it agrees sufficiently with our own habits. But therein lies the very essence of human history. Other living things have no history, or at any rate none that they can remember and record ; whereas Man is differentiated from them by nothing else so much as by his power of accumulating a social tradition, and using it to enlarge and reinforce individual experience as a means of survival. Now, the difficulties of thinking at once analytically and on a very concrete basis being what they are, it is likely that amplitude of scope and

43

inconclusiveness of method must ever go together. Even so, we claim to be an organization, not a mob. To a like extent, therefore, we are pledged to pursue some common policy ; which can be done only if we constantly re-examine the further implications of the task that lies before us.

The point, then, that I propose for consideration is the following : Does Sociology set out to treat human values as if they were simply facts ; and, if so, how far is such a treatment feasible and useful ?

At the risk of divagating into metaphysics, let us begin by seeking a meaning for " fact " as the empirical or positive sciences understand the term. Needless to say, the historical associations of Sociology are with the sciences in question. Nay, its very name—that " convenient barbarism " as Mill called it—proves as much ; for it could never have been invented by a friend of the classics and therefore an ally of the humanities. Anyone who has been brought up on the humanities—the typical Oxford man, for instance—will, long before he ever heard of Sociology, have become familiar with a social science modelled on the politics of Plato and Aristotle. Now there we have, as all must admit, an intellectual discipline of superb range ; more especially since, for the Greek, politics not only carried on a joint business with ethics, but even ranked as the senior partner of the firm. These sciences of the ancients, however, were essentially normative rather than inductive in method. They might by way of inchoate induction review current opinions in some detail before proceeding to prescribe ideal principles conformable with human aspiration ; but ultimately the whole constructive argument hung on certain values laid down legislator-fashion. Such values might, or might not, be stated in an absolute form, but the absoluteness in question could only be analogous to that of one professing to rule by divine right, a pope claiming infallibility. They were necessary only so far as an act of pure will could make them so.

In sharp contrast we have the sciences of the inductive type which base their logical constructions not on values but on facts. It is in such a company that Sociology would have itself enrolled. And yet it must be confessed that its position in the hierarchy is never likely to be high. It is not of such stuff that Presidents of the Royal Society are made. Only in the normative sphere does social science come to its own as the royal science *par excellence*, the supreme concern of the philosopher-king. The trouble is, from the inductive point of view, that social facts regarded as building material do not give good enough results ; they seem to produce nothing stable. For the sciences of this order pride themselves on the solidity of their foundations. They proclaim themselves positive, being apparently unaware that the word " positive " is ambiguous, and might just as legitimately be used to connote normativeness. Thus the judicious Hooker writes : " In laws, that which is natural bindeth universally ; that which is positive, not so." But the inductive sciences insist in the same breath that they are both natural and positive. Perhaps it is as well that they should be allowed to take over an expression that will serve to remind them that in selecting fact as their foundation they are positing something too. They put their trust in the uniformities of sensible observation, just as the normative sciences do in the uniformities of moral conviction. Now it would take a lot to persuade an anthropologist accustomed to survey at large the manifold vagaries of insensate humanity that strict uniformity, affording an evidential ground for a presumption of necessary law, is to be reached in either of these directions ; nor, as any psychologist must allow, is individual experience less prone to vacillation, verging on self-contradiction, in its various impressions and affirmations. Nevertheless, it may well be true as a working rule of life that, relatively speaking, judgments of fact are to be reckoned more substantial as a groundwork of intellectual castle-building than judgments of value, if only because they are more widely shared, and hence offer better support for the average gregarious man. Tried by that most ultimate of practical standards, mass-opinion, the democratic vote, the argument from consent, sense is more steadfast than sensibility. Thus it makes an

45

excellent compass for setting a course. It is well to remember, however, that this in itself can never amount to planning the voyage ; for in the latter case the owner's wishes are paramount, and the mechanics of the enterprise become secondary to its guiding purpose.

Fact, then, is fate, the last word of the sense-world, the ineluctable aspect of experience, which is hypostatized by the uncritical mind, so that it seems to possess, if not a will of its own, at any rate a self-dependent power of cold obstruction limiting our will from without. It is thus the converse—one might almost say the adverse—of spontaneity, either as we know it in ourselves or as we extend it by analogy to living things in general. Literally and etymologically fact stands for the inevitability of the already accomplished, the irreversibility of a has-been, the finality of a " that is that." The Latin *factum* is not to be translated " made " in this connexion ; for it is precisely our ultimate participation in the matter that is conveniently ignored. It means, on the contrary, " done " and, so to speak, " done with " ; because the only way to get level with the past and have a chance of undoing it would be to abolish time altogether. If, on the other hand, the notion of fact is projected into the future, or even if it is treated as a condition of the fleeting yet ever-immediate present, time no longer conspires with the past to make it seemingly all-compelling as a *vis a tergo*. Actual or eventual fact is always contingent, and always qualified by present or impending novelty. However thoroughly we may try to resolve the already given into a complex of would-be timeless causes operating with a like unalterable givenness, our experience on that side of it which is intuitively aware of a forward thrust, an *élan vital*, reveals a generative vigour which is *pro tanto* fact-transforming and fate-defying. Moreover, the event is wont to confirm this intimation of the human spirit by its truly dramatic unexpectedness ; for to say that " the more it changes, the more it is the same thing " is but one of those paradoxes that make play with the lesser half-truth. That we find ourselves alive and active in a world that has to put up with it is retrospectively a fact.

But immediately and therefore in a deeper sense it is likewise
a challenge to fact, a declaration of independence validating
our right to struggle for an existence of our own conditioning
and choosing.

It only remains to add that the sciences which deal with
fact are not necessarily in the pay of the enemy, as too warm
adherents of the humanistic faith might sometimes seem to
imagine. They are rather spies sent out to discover his real
strength and weakness. They advise us, for instance, not to
attack him in his outlying dominions such as the extra-
galactic systems, but to look for chances nearer home, namely
in the more hospitable parts of this planet—a poor thing,
perhaps, but already on the way to become our own. For
we need sound information, and cool calculation based
thereon, as well as sheer buoyancy of enthusiasm, if we are
to get the better of fact in the sense of the sum total of the
difficulties of the vital situation, not excluding our own former
shortcomings and mistakes. It is, however, on the latter
subject that our intelligence-service tends to be especially
weak. The reason is plain enough, namely, that to think of
ourselves in terms of fact is, in view of our mixed record, apt
to prove decidedly damaging to our self-conceit. Hence,
whereas we rejoice in objectivity or matter-of-factness, so
long as material conditions are alone in question, the most
accurate observations concerning our rather rickety morale,
however needful it may be to take account of its fluctuations
in a war so prolonged and with a battle-front so far-flung,
meet with a doubtful reception at headquarters, and as
likely as not are quickly shelved. To drop the fighting
metaphor—though truly the struggle for existence is no
unfair description of the actualities of the life-process—a
scientific in the sense of an objective history of mankind is not
readily forthcoming, and chiefly, one may suspect, because
nobody wants to have his past brought up against him.
Even if science claims as its own that so-called child of
Nature, the savage, as standing at one clear remove from the
practical interests of the dominant peoples, it cannot wholly
prevent him from being converted by tiresome if well-

meaning persons into matter for edification, whether as a shining example or as an awful warning. So, whereas the historian-artist, if he has the flattering touch, can get good money out of the snobs and launch out as a gentleman, the sociologist, as a historian-photographer whose talents lean towards the mechanical, finds but little lucrative employment, and must be content to set up in a back-street in the trading part of the town. Yet, putting aside the question of remuneration as affording no sure index of merit, we have to enquire how far two distinct and in some sense rival modes of depicting the true Man, outside and inside, countenance and character in one, can exist side by side or even be combined ; so that the worker in values and the worker in facts, the artist and the artificer, can both in their several ways help us to envisage ourselves as nearly to the life as may be possible in the circumstances.

To turn now to the subject of value, this clearly relates to life potential rather than to life as simply lived up to the present. A value can be described as an imagined satisfaction, if we look at it from the side of feeling ; while, regarded from the side of the will, it is an imagined option. No human being can live for the moment so utterly as not to be partly possessed by hopes and fears relating to a beyond ; and this he cannot refrain from translating into terms of possible experience, even when imminent death stares him in the face. On the other hand, no calculation based on previous experience can enable him to take the complete measure of what lies ahead ; nor at a pinch does the wise man put his trust in any mere knowledge so much as in a native resolution which seems to represent the life-force itself, since it bids him " never say die." This power may in certain conditions compel him, against all likelihood of safety or success, to play the part of an *Athanasius contra mundum ;* that is, for the sake of some hitherto unrealized principle, to fly in the face of all known facts, as even the most learned men are privileged to know them. Thereupon, it may be, the single seed proves a mutation destined to cover whole continents with unexpected harvests as by a miracle of special creation. Not that the pioneer-temper is universal or even common among mankind. Unfortunately, as the Preacher reminds us, the battle is not to the strong, or at least not always ; whereas, whichever way it goes, certain human jackals take their pickings, and after their sneaking fashion also manage to inherit the earth. Nevertheless, the leadership of the world is vested in its heroes, who are likewise visionaries to a man. These know, or perhaps rather feel, that the core of reality is constituted not by fact but by inspiration.

It is not, however, with the man of action and his deeds that we are concerned here, but rather with the kind of thought that can make his purposes explicit, whether to him or to the world that has need of him. A value always embodies a meaning such as can be expressed in rational terms ; though it must not be forgotten that rationality comprises more than mere understanding, and makes its appeal to head and heart alike and together. When we speak of the normative disciplines as sciences—arts being at least as appropriate a description of them—we may easily be led to over-emphasize their purely intellectual side. No doubt their typical form is that of a ratiocinative system which is exhibited in its dependence on some end ; and often this end is taken to be so obviously desirable that very little is said to advertise its attractions, as it were on the principle that good wine needs no bush. Hence a pettifogging mind is apt to ignore the *proœmium legis* which defines the statutory intention as a whole, and to pass on to wrestle with the thousand and one consequential regulations that swell the body of the act. Experience assures us, however, that the minor tactics of life hardly affect the main issue if only the strategy is sound. Though it may require the meticulous expertness of an Aristotle to codify the maxims of a reasonable ethics, it takes a Plato, burning with the fire of prophecy, to announce the Idea of the Good as the crowning justification of all human striving. Let the myriad applications of it fail, yet the golden rule stands inviolate, being ever beyond our grasp and yet within our reach. To cite the Preacher once more : " Lo this only have I found, that God hath made man upright ; but they have sought out many inventions."

The real strength, then, of the normative treatment of value in any form, supreme or departmental, lies, or ought to lie, in its initial affirmation ; for it is constitutive and so to speak creative of all that follows. This, however, is work for a philosopher-poet rather than for a man of science, for the pulpit rather than for the laboratory. It calls for the fervour that can propagate as well as promulgate a creed, and not simply for the canniness that estimates the profits and risks thereby entailed. It must declare the authority of the director, over and above the mere calculations of the accountant. To draw up a prospectus as compared with a mere balance-sheet must ever rank as the more responsible task, because, while brains count for at least as much, character becomes of far greater importance. Nevertheless, while we thus assign to the normative exercise of the mind the higher part, because in championing value it brings Man's stock of faculties more fully into play, we must none the less allow that it is extremely difficult in practice to attain to the level of reason, or whatever we are to call the outcome of a good working-alliance between intellect and emotion. To blow hot and cold together is proverbially unmanageable, and yet a cool disinterestedness has somehow to be combined with an interest as warm as the very blood in our veins. Yet the thing can be done ; for Plato at least does it, standing therefore in my eyes, if I may make a personal confession, as the completest and most divine of philosophers—I had almost said " and of anthropologists as well." There are some of my colleagues, however, in the last-named depart-ment of science who hold that their first business is to dehumanize Man as far as it may be feasible. One must play onlooker, they say, to the game of life and pretend that one is no player oneself, or at any rate a retired one, and with not one farthing staked on the result. Such a spectator, like some visitor from the Celestial Empire witnessing the strange antics of sundry foreign devils, might occasionally spy out details apt to be overlooked or misinterpreted by the more enthusiastic type of partisan. On the other hand, to appreciate a sporting event for what it is truly worth to those more immediately concerned might, nay, almost certainly would, be clean beyond his range. Thus there are difficulties either way in taking stock of the human life-struggle, whether one is too sympathetic to be judge of its incidents fairly, or not sympathetic enough to be in touch with its spirit. A normative interpretation, however, must show sympathy at all costs. It must render an account of some vital purpose in terms of its own self-justifying conviction, its esoteric profession of faith, its conscience. The gist of a gospel must ever be in its inward appeal.

Here, then, we have one, and indeed the chief, reason why we are bound to discriminate carefully between the kind of science that attempts to vindicate a value, and the kind that simply deals with it as a fact, or, in other words, sets out to explicate its historic content. While the latter is but an intellectual exercise, the former involves the combined use of intellect and feeling. Hence, whether he is fully conscious of it or not, every normative thinker is at heart a mystic. What the formalists call realism, meaning mere phenomenalism, will not suffice him when he seeks to portray the ideal as a real presence manifesting itself as such to those who surrender to its activating influence and yet can find it only in the act of continuing to seek it. Such a type of thinking cannot but be in the closest alliance with the life of action ; though the man of action is not to be identified with the hero of the history-books, the noisy man as one might say, but rather with any humble individual who tries to live the good life with all his might. The latter can, however, but do this according to his lights, and must therefore treat it as part of his duty that he be not deficient in illumination through any fault of his own. Here, then, in the supplying of such educative means as may be needed, lies the opportunity of the teacher of outstanding intellectual power, anyone, that is, who, whether ranking professionally as philosopher or preacher, poet or novelist, is a serious-minded and well-

informed exponent of the art of good living. His special
business is to strengthen the plain man's natural desire for
good by directing it towards the more abiding satisfactions.
Unless the bias were already there, nothing could be done
to help ; so that the trainer must, as it were, be leaning over
the shoulder of the actual player when he shows him how to
improve his aim. If, on the other hand, he were merely to
stand aloof and cry " bad shot ! " it is doubtful if he would
succeed in being more than a spoil-sport. Hence, though
criticism must enter into any examination of value on normative lines, it has at the same time to be a friendly criticism,
because expressive of a supreme interest shared in common.
Thus the Socratic method as applied to morals might seem
on the face of it negative. Yet the purely zetetic side of the
argument is throughout secondary to a dogmatic assurance
that true wisdom will be found equivalent to complete
goodness. Socrates always says in effect to his companion :
" You and I are clearly at one in wanting the Good, so let
us reason it out together what it is precisely that we do
want."
Contrast any treatment of fact as such. It involves no
affirmation of value at all except the trifling one, appealing
primarily to the curiosity-hunter, that any curious bit of
information is somehow worth picking up because at least
it is always news. Strictly speaking, from the standpoint of
pure science even the pragmatist contention that any piece
of knowledge is bound to pay in the long run is not a relevant
consideration in the study of fact for its own sake. On the
other hand, this parsimony in the matter of its claim to throw
a direct light on the desirable aspect of things renders the
mind wonderfully cool in its bearings. The intellectual
machine runs as smoothly as ever when at the bidding of
the astronomers we contemplate the fading out of Man with
the rest of the so-called biosphere from the face of an effete
planet. Like certain bacilli, such notions are not killed off
when they approach the zero-point of a moral temperature
needing greatly to be raised before we can rejoice in being
alive. Now some facts undoubtedly lend themselves better

to this attitude of frigid detachment than do others. Inorganic matter, for instance, is further away from us than
any thing organic, because, however addicted we may be to
what Mr. Ruskin calls the pathetic fallacy, brute nature
must ever baffle any overtures in the way of intimacy on
our part :

> and still thought and mind
> Will hurry us with them on their homeless march,
> Over the unallied, unopening earth,
> Over the unrecognising sea.

Since, however, fact as such implies the deadness of the
fulfilled event, it ought to be theoretically possible to view
anything given in history with the same indifference. Yet,
as a matter of actual experience, it is far harder. Cæsar may
be dust, but Cæsarism is with us still. Even Neanderthal Man,
a being of possibly another species, had ritual practices
implying some belief in a life after death such as declare him
a spiritual brother, and hence one to be regarded none too
impartially but rather as we regard a backward member of
the family. Nevertheless, the judgment of fact, in whatever
application it is found at its best, will serve as the exemplar
of as perfect an exteriorization of thought as is humanly
attainable. It suggests a method whereby objectification can
be effected in a degree incompatible with the effective
exegesis of any teleological system. If we symbolize the
indissoluble relation of subject and object by the combined
letters SO, then, whereas big S and little o may stand for a
normative truth, one that embodies mere matter-of-factness
qualifies a large O with s as small as imagination will tolerate.
By taking, then, the so-called positive sciences for our model,
we sociologists are expressly adopting an exterior and, so to
speak, unmoral attitude towards our subject. " No political
axes ground here " is the notice over our door. If we
would be social reformers as well, we must develop a double
personality. Nay, the more complete the dissociation
between our two spheres of interest, the better sociologists
we shall be ; though not necessarily the better men, unless
with the aid of philosophy or religion we can afterwards

pass on to a higher plane of thought. Thereupon one can take in the drama of life in the meaning it has, not for the mere scene-shifter, but for the actor as interpreter of the author's purpose. Let the sociologist, then, make a merit of being able to keep cool; but it is only fair to remember that working in company with the out-of-doors staff is a chilly proceeding anyhow, and that his spiritual health may suffer unless he warms himself at the home-fire at the end of the day.

So far we have been dealing with a fairly conspicuous ground of disparity as between a valuation *per se* and a mere statement of fact about a valuation, consisting in the respective presence and absence of a practical interest, so that in the one case there can be friendliness and in the other neutrality. The next point of difference is, however, not so obvious, yet of sufficient importance to be worth more attention than, so far as I know, it has hitherto received. It is shortly this—that a normative treatment must be singly determined because it exhibits means in their due subordination to some all-embracing end ; whereas a treatment of facts in their diversity inevitably allows simplification to halt a long way short of unification. In other words, the one method professes to be absolute, the other is satisfied with relativity. The moralist who can assign no positive content to the good is soon convicted of having chosen the wrong vocation. A historian of morals, on the other hand, can provide a variety-entertainment without scandal. Now this obligation on the part of the teleological thinker to present his ideal in some one authentic shape and subject to no Ovidian metamorphoses is bound, in practice, to make for a certain narrowness. Normative systems rapidly become old-fashioned. Again, they do not amalgamate or grow out of one another, in marked contrast with the inductive sciences which welcome innovations as so much increment, and by so doing can maintain a continuous expansion. Granted that, in a highly conservative state of society, it might be deemed superfluous to issue more than one edition of the law :

> But no ! they rubb'd through yesterday
> In their hereditary way,
> And they will rub through, if they can,
> To-morrow on the self-same plan.

In a word, the static state is necessarily orthodox, the two conditions implying one another. When, however, we turn to a society such as our own, which can be termed dynamic, in the sense that it is not in equilibrium but in constant motion, whether this motion be but sheer change or amount to a genuine progress, then, to borrow from the artist his description of a well-known device for imparting vigour to his composition, the statesman must likewise learn " to keep it loose." To change the metaphor a little, the social reformer of to-day must not screw up the parts of his machine too tight, or, under the high power furnished by civilization, it will shake itself to pieces. Professor Bouglé, in his interesting *Leçons de Sociologie sur l'Evolution des Valeurs* (Paris, 1929, ch. V), makes what he calls " polytelism " the distinguishing mark of the modern community, no doubt thinking more especially of the type that is democratic and liberal. On the other hand, certain recent developments in Europe, whether they are to be regarded as reactionary or not, certainly place solidarity before liberty, even in the form of liberty of thought, that first and most precious article in the charter of a true civilization. Meanwhile, in this country at least, we know by experience that it is possible to agree to differ. Just as in private life we try to cultivate a sense of humour, so in politics we favour a spirit of toleration and compromise which, thanks to the individuality generated in the process, brings about intelligent co-operation, and leaves those who have the parade-ground habit of mind in a perpetual state of wonder how it is that such gay and giddy irregulars should so consistently manage to " muddle through." Nay, even in the sphere of religion, where, if anywhere, unanimity might be insisted on as the only way of making good a claim to catholicity, the multitudinous Protestant communions pursue their several paths without much effort to keep in touch, yet for all that with an unconscious sense of direction that allows a common advance ; while even within a particular body of worshippers schism is rare just because latitudinarianism is so general. In short, what Bagehot has termed the "Age of Discussion" has dawned for mankind, so as to release him from that moral servitude which is exemplified by typical savagery. Under a régime of taboo a man cannot call his soul his own ; and perhaps in the circumstances he is better off if he does not know that a soul is included in his private possessions. Fortunately or unfortunately, however, the civilized man is no longer in this state of ignorance. Having once tasted of the tree of the knowledge of good and evil, he cannot but recognize a grovelling subjection to some external sanction, however speciously decked out with the insignia of authority, as an evil thing, the very negation of all self-development and true manhood. Thus we may suspect the wolf in sheep's clothing when we are bidden to acquiesce in so mutton-headed a doctrine of the higher possiblities of the social life. Having fought our way out of mere gregariousness, we are not going to be forced back into it without a protest, or, if it comes to it, a fight.

Yet any liberalism must always prove a stumbling-block to the normative thinker, because his treatment of the social and moral problem is bound by the logic of the situation to be authoritative through and through. His kind of truth cannot be proclaimed simultaneously by rival speakers aloft on their separate tubs. Whatever discords his instruments might produce if left to themselves, his special function is to extract from them some sort of concord. The chances are of course that, of all the sounds that nature and art between them can offer, his selection can utilize but a few. Nay, in the composition of such a symphony varying degrees of richness may be sought. The idealist attracted by the simple life might embody his conception of it in a Bach fugue; whereas a Beethoven sonata might be more appropriate if justice must be done to the complexities of the city which, as Plato puts it, lives delicately. Meanwhile, in any case, the given elements, scant or abundant, must be harmonized, or else the artist has failed in his purpose. Moreover, tastes differ ; so that the next musician will assuredly weave his notes together differently.

How, then, is the " polytelism " apparently so vital to the existence of the liberal type of modern society to co-exist with this " monotelistic " bent of the normative mind. Here, as it seems to me, is the great opportunity of Sociology. Studying the facts of the social life as it does without attempting to reconcile their inconsistencies, as these must appear to one who is called upon to choose between them, it testifies to the great variety of the possible satisfactions hitherto accessible to Man in the course of his long career ; though it always remains for the normative thinker to convert them into live options at his own discretion. On the other hand, I would protest with all my might against Durkheim's view that social facts as such can be " coercive." Such a deterministic interpretation of the pressure that they undoubtedly exercise on human wills, alike in their united capacity and as severally regarded, is a sheer piece of exaggeration more suited for a rhetorical than for a scientific context. We are not obliged by some iron necessity to abide by the language, the religion, the political system, into which we happen to have been born. However ready-made our convictions may seem to be, there is consent, tacit or explicit, at the very root of them, and hence also the possibility of dissent. Only an outmoded psychology, the associationism of a James Mill, might seek to explain any one of us as the passive product of his social environment. Nay, it would be more reasonable to fly to the other extreme, and to attribute a certain measure of autonomy to every living thing. For in its small way the merest bacillus makes a bid for the existence that suits it best ; and, though our intelligence may look down on its instinct, we do not always get the better of it in that essentially competitive larger society to which we both belong. It remains, then, for Sociology to present its social facts for what they are, namely given conditions that at most have to be treated with more or less respect by those who would walk cautiously. These need to be as fully acquainted as possible with both the advantages and the dangers hitherto attendant on this or that line of advance ; though at the same time the thought should never be out of their minds that the obstacle of yesterday may be so re-adapted as to provide the bridge of to-morrow.

To put the matter in a different way, Sociology makes no affirmations about values, but so presents facts that they can be used for the criticism of values. Ideality by itself is not enough to commend a course of conduct in the eyes of those who have to make the best of this present world. Possibility must also be considered. Choice, though it always involves risk, need not be blind. The sociologist, however, leaves the estimation of his results in terms of choice to others, namely, the framers of ideals. His duty ends when he has surveyed the world from China to Peru, and shown how all sorts and conditions of men, in relation to social environments correspondingly various, have found it possible or impossible to live at all. He is not even required to show whether they have lived enjoyably in the given circumstances ; because at that point there arises the question of what makes life worth living—a matter involving value-judgments not likely to be uniform the whole world over. These very judgments, however, in all their heterogeneity become facts as soon as they have taken effect, and so can be swept into the net of the sociologist as a most instructive part of his material. He can show, for instance, that this people opted for agriculture ; that another preferred the pastoral life ; that a third developed a taste for trade. In their subsequent histories, which it is the part of the sociologist to expound in the light of these leanings, there plainly lies a moral. Yet it is not for him to draw this, beyond stating that as a fact some prospered, while others were wiped out.

Thus the sociologist presides over a bazaar of social experiments, open to all customers to buy or not to buy as they please ; his business being simply to see to it that the goods displayed are so labelled that bankrupt stock can be distinguished from the products of some leading house. Now this same bazaar-metaphor was applied by Plato to a democracy, because he discerned in that type of polity a polytelism of which, not having had the advantage of studying M. Bouglé, he thoroughly disapproved. But Plato as the most authoritarian of idealists, though he does not burke the problem of the possibility of his ideal state as something to be established firmly on earth and not merely set up as a pattern in heaven, does not find it any too easy to show how passive obedience on the part of the proletariat is to be conjoined with a liberal education for the ruling classes ; nor indeed is it apparent how a training on mathematical lines is to fit the legislator for dealing with the contingent. In any case Plato thought in hundreds, where we have to think in thousands and millions, of individual citizens, these having not only duties but rights, and more especially the right to self-determination. Let these countless numbers, then, be required to wear uniform on state occasions, if the authorities will have it so ; though the assumption is a little odd that patriotic hearts have to be worn on the sleeve. Even so, there is surely room also for sartorial licence in private life, even at times if it verges on eccentricity ; which, if by no means the equivalent of individuality, is at least the inseparable accident of the latter and hence symptomatic of its presence. In our sociological mart, then, where the purchaser is offered a variegated display of modes, ranging from genuine antiques to the latest novelty, while the choice rests entirely with himself, there are all the makings of a liberal education. At the same time we must not forget that it needs an educator to show us how to use these makings. If therefore the sociologist were to overstep his part and force his goods upon a gullible public, then indeed he would be in the employ of Beelzebub, the Chief Lord of Vanity Fair. Yet, as Bunyan allows, " the Way to the Caelestial City lies just through this Town." It is an inevitable incident in the pilgrim's discipline and progress that he should learn to rate sham merchandise at its true worth, and, though all the kingdoms of the world be shown him by way of temptation, should continue to gaze steadfastly ahead towards the journey's end. So much more noble, then, is the function, and so much greater the responsibility, of the thinker who would guide human choice than those of the historical student who merely provides chcice with its materials, the bad with the good. Thus social facts, *pace* Durkheim, are not coercive. That can only be the attribute of a value to which we have, of our own freedom, vowed an absolute devotion. External nature stands for opportunity, never for sheer necessity ; and this holds true even of human nature so far as it can be externalized as fact, because it has ceased to be, and has become but the heritage of the past. Coercive fact, then, like fate, is a bogey which can but dog the coward whose eyes are in his back. Internally viewed, moral evolution is rather a progress along the line of greatest resistance !

These considerations touching the relation of fact and value, and the types of social study that severally assign the precedence to these diverse yet ultimately complementary principles, might be extended to cover more points of difference ; while as much again might be said by way of reconciling the two points of view. But I have thought it better to emphasize a few plain truths, if indeed they are as plain and true as I take them to be ; trusting that nothing I have said will offend my fellow-sociologists, even if I may have occasionally seemed inclined to clip their wings. Yet, after all, we are *non solum Angeli sed etiam Angli*. Self-depreciation positively suits, I might almost say flatters, our national humour. It can do us no harm to remind ourselves that it is not for us as sociologists to lay down the law about how society should behave ; that so long as we are dealing with " is " we must try to leave " ought " entirely out of account. Cold comfort this, it will be said, for those of us who are likewise social workers. Not at all. Even the man of science can be a human being in his off-hours. It is simply a question of practising suspense of judgment in the form of the predication of a value, or in other words, the profession of a creed, so long as we are engaged in determining simply whether certain things have happened, and so, in proportion to the constancy of their recurrence, are likely to happen again. Sooner or later, however, the confessional attitude can be resumed with perfect propriety, and facts already duly attested can be re-examined and re-interpreted in the light of their relevance to the cause embraced. Any well-trained mind can be organized on the model of a court in which the functions of the witness-box and the advocate's bench are kept apart for the better enlightenment of the supreme faculty which delivers final judgment. Associated though we are with the votaries of the natural sciences, we sociologists can perhaps never hope to equal the magnificent impartiality of the experts whose evidence relates to the purely technical aspects of the case. For the party on his trial is Man, and, since therefore it is from the dock that we ourselves step across into the witness-box, we are inevitably testifying on the side of the defence, strive we never so hard to be honest. Yet stern Justice abominates perjury, and even tender Mercy would have the whole truth before it forgives. So let it be our special task to lay bare the family history in all its details, whether reputable or the reverse. The rest is for the judge and the court missionary to decide between them.

3. RACE AND SOCIETY

Every student of human nature as it is revealed in history must organize his work on the basis of a time-scheme. Unless he has first arranged his facts in their given relations of simultaneity and succession, he will inevitably produce mythology, not science. Figuratively speaking, then, he needs a clock ; and I am going to suggest that such a clock can be constructed after the usual manner with two hands, the one to tell the hours and the other the minutes. How Race and Society are severally to provide the hour-hand and the minute-hand of such a secular time-piece will have to be explained at some length ; but in the meantime this metaphor of the clock may help to keep before the mind's eye my main contention, namely, that in considering human development in relation to the time-process we can distinguish in racial change and social change respectively a slower and a quicker rate of movement. For methodological purposes, then, I propose the joint use of such a pair of markers for measuring the rate of that sequence of related events which we can construe as an evolution or as a progress as we choose.

It is, perhaps, not irrelevant to add that, if we are thinking in terms of geological or biological time as a whole, our anthropological clock would need to be wound up for but a single day in order to cover the life-history of our species ; and that though we construe the notion of species ever so liberally, and, in defiance of the systematists, bring *Homo sapiens* and *Homo vix sapiens* under one and the same liberal category. On the other hand, the clock must run for a complete year of such days if justice is to be done to the longevity of Lingula, which for untold millions of years has touched perfection as embodied in the perfect stick-in-the-mud—a fundamentalist, therefore, who has the best of excuses for treating Evolution as if it did not exist.

To revert to this alleged need of a two-fold standard of reference in our historical time-reckoning, it is possible to offer good psychological reason for such a duplication of methods. For our intelligence on its speculative side aims at being long-sighted. In its practical capacity, however, it must at all costs be sharp-sighted. Thus, when our object is simply to explore the universe at large, we are unwilling to stop short of its uttermost dimensions, having learnt like Alice to expand or contract at will, so as either to strike the ceiling of the extra-galactic or pass through the keyhole of the ultra-microscopic. Yet there are other times when such cosmic interests must give way to the claims of domestic affairs, comprised as these primarily are within a social present having an overlap of roughly three generations, say, the range of a century or something less. However difficult the history of our own times may be to grasp objectively, it is of vital concern for us to try to understand its true meaning and tendency, seeing that we live and move and have our being in it. In this narrower but more immediate context we must rely on intensive rather than extensive thinking, since, whatever we may conclude as philosophers about the ultimate whence or whither of mankind, it remains our first duty to keep it going here and now. The long day's work, then, may be spaced out in hours, but the pressing task in hand must be regulated to the minute.

Now if I am right in supposing that racial questions and social questions can be to a large extent kept apart because the major and minor types of adaptive process with which they have severally to do involve an entirely different timing, it might seem to follow that distinct bodies of specialists should each undertake to watch one kind of movement rather than the other. Without insisting on too rigid a division of labour in a subject which touches us all too closely to become fully amenable to the cold-blooded dispositions of the pure theorist, I would suggest to the sociologist that he claim the topic of Society for his own, leaving the complementary, yet as it were remoter, topic of Race to the anthropologist; for whom Sociology, or as he prefers to term it Social Anthropology, is not so much an end in itself as a means to the solution of a larger problem, namely the discovery of Man's place in the general scheme of living Nature. In thus demarcating what I believe to be natural frontiers between allied disciplines, I do not forget that both are ultimately seeking to take stock of the same time-series of occurrences, for all that it appears to move slowly or quickly according as one judges the pace by one kind of observable change or by the other. Neither kind of change, of course, reveals itself at a first glance; and just as in the clock of our metaphor the eye detects no continuous progression, whether of the hour-hand or for the most part even of the more nimble minute-hand, but in each case the movement has to be inferred by means of a mind-picture of contrasted positions more or less static in themselves, so it is only by the aid of the Comparative Method that we can assure ourselves that Race and Society are both of them alike in ceaseless transformation, though each after its own characteristic fashion. To different specialists, then, I would assign the distinct and separable tasks of devising a standard of measurement appropriate for the study of each of these two processes, the one so deliberate, the other relatively so rapid. My advice to the sociologist is to concentrate on Society, and to abandon Race to his anthropological colleague, since, as I shall try to show, the system of short reckoning required in the one connexion is not easily combined with the system of long reckoning demanded in the other ; and, for the rest, it is sometimes convenient to be able to refer awkward questions to the other partner.

To begin, then, with the subject of Race, let us avoid loose and popular misapplications of the term, and assign it its strict biological meaning. In this technical sense it stands simply for the congenital element in any organism as distinguished from that experiential or acquired element which is no less necessary to the vital make-up of every living thing. Now this partition into elements is, no doubt, in the first instance a logical artifice ; and, so far as it is this and no more, Race would have to be regarded as something wholly abstract, nay, as Topinard in his day was disposed to think, something so abstract as almost to be transcendental. If, however, we accept modern opinion, which on the whole adheres to Weismann's hard-and-fast distinction between germ-plasm and body-plasm, our abstraction takes on a concrete form ; for we can now identify Race with just that continuum of preformed yet plastic vital energy which the germ-plasm incorporates and passes on. Meanwhile, the geneticist, with the help of his chromosomes and genes, has subjected the notion, or we can at this point fairly say the fact, of this vehicular mode of transmission on the part of the inheritable characters to further analysis ; so that we can at length follow in remarkable detail the workings of those sole agents of the reproductive process, the gametes, as in the course of the tireless ballet of sex they clasp and again unclasp so as in altered guise to find fresh partners. As for the individual whose very existence depends on one of these temporary partnerships, it is by no means clear that in himself and as such he plays more than at most a subordinate part in the shaping of Race. He can be resolved into a community of cells. A few of these retain the race-carrying function—royal cells, as we may call them ; but the rest are degraded into neuters, mere slaves, as it were, of the economic machine. These subsidiary parts of the organism constituting the body-plasm have indeed an indirect bearing on generation in so far as without their support the special vehicles of the germ-plasm would lack physical sustenance and die. Yet this assistance is quasi-external, in that the specialization of the royal cell enables it to assimilate only what is consistent with the maintenance of its inherent bias. It remains qualitatively true to itself ; and the nature of the bio-chemical fare on which it subsists at the hands of the ministrant cells makes no difference to the fulfilment of its biological purpose, unless possibly to endow it with a greater or lesser degree of dynamic vigour. All this, of course, is highly theoretical, but it may at least be argued that the available evidence points in this direction.

Accordingly it is within the germ-plasm itself rather than in the organism as a whole that we must seek for the sources of the special kind of change manifested in the propagation of racial characters and types. It is usually termed variability, as contrasted with the modifiability that goes with individual experience. Such variability includes the case of the mutation ; which expression simply implies that the observed change is a big one—a decisive shift-over of the racial kaleidoscope to a new pattern. Now germinal variation of a degree entitling it to rank as mutation has been produced experimentally by means of X-rays ; and what can thus be done in the laboratory is no doubt accomplished on a larger scale by Nature, using cosmic rays as her instruments. It by no means follows, however, that any such change is for the better. There could be no greater fallacy than to regard the mutation offhand as an improvement introduced by a Providence who with unfailing beneficence furthers the cause of progress. On the contrary, the average mutation would seem to reveal a weakness, to make for inferiority ; as if a stable equilibrium of forces, a sort of pre-established harmony, had been disturbed in vain, being so well poised as spontaneously to right itself in the long run. Indeed, the phenomena of hybridization tell the same story of true-breeding forms that, whether dominant or recessive, reassert themselves, even though forced to blend for a time, and suffer a confusion of characters, a lapse into the nondescript. Meanwhile, this fidelity to type is never so absolute as not to permit of deviations from the average in all directions, yet so that there occurs no noticeable alteration in the mean. Thus a family face or a family disposition may come out in a number of persons otherwise markedly divergent in their traits. Such considerations, then, taken together, make it likely that the germ-plasm is as it were resistant to such shocks, external or even internal, as threaten to throw it out of its accustomed stride. To adopt the language of politics, one may suspect it of being strictly constitutionalist in its attitude towards any reform of a drastic kind, so that whenever change comes at last, as must happen in a changing world, then as when a tadpole is metamorphosed into a frog it comes by revolution. It is certainly provocative of thought along these lines that the biological record should provide us with what may fairly count as species in such multitudes, whereas the intermediate forms, which ought to occur if evolution depended on an accumulation of small differences, are mostly missing. That record is admittedly imperfect ; yet it would fit the known facts well enough to conceive evolution as a sharply graduated advance, a series of rather sudden transitions between steps slow enough to favour the recovery of balance at whatever cost to momentum.

At this point it may be objected that it is premature to discuss the nature of the evolutionary process when variation alone has been taken into acocunt, and the complementary principle of natural selection allowed no say in the matter. What Weismann, however, described as " the all-sufficiency of natural selection " was meant by him to convey no more than that, given a sufficiency of variations in the germ-plasm, Nature, unassisted, would thereupon see to it that the failures were duly weeded out. Thus the initiative is all along conceded to Race, while Environment, deprived of any Lamarckian *vis formativa*, is granted the bare right of serving the strongest master. Passive condition as it is, however, it responds but grudgingly to exploitation : and life clings precariously to the surface of a revolving globe which, so far as we can prophesy, might at any time fall back into the inhospitable habits prevailing generally in planetary and stellar circles. Still here we are, myriads of us alive and kicking, from the bacteria upwards to Man ; and, thanks to unlimited competition, effective ways of living, very largely at the expense of one another, have been continuously devised. Thereby the life-force—the mysterious unity immanent in all this plurality—has kept abreast of climatic and geographical changes, nay, in a sense a little ahead of them, since logically the experimental impulse must be there before any kind of adaption can be so much as attempted.

In saying this, however, I must not appear for one moment to underrate the power of the environment ; for, negative and purely eliminatory power though it be, it acts as a penal sanction whereby the aberrations of the all too riotous life-principle are sternly held in check. And, like the penal sanction of the legislator, it is apt to be slow, though none the less sure, in making itself felt. Both in the animal and vegetable kingdoms there are times when justice appears to sleep—when delinquent variations seem to enjoy a thieves' holiday without heed of future reckoning. But, as the Psalmist declares, " when all the workers of iniquity do flourish, it is that they shall be destroyed for ever." I hardly know if it would be in point to refer in this context to the fate which overwhelmed the Trilobites or the Dinosaurs ; which, even if the environment, physical and biological, proved too much for them in the end, at any rate luxuriated in all the privileges of aristocrats during the æons through which they lasted. I am thinking rather of the havoc which one hard winter, coming on the heels of a succession of mild ones, will cause among the tenderer species of our birds. When the pinch comes, what a massacre of lives that would have no difficulty in coping with normal conditions ! Yet those tougher specimens which actually succeed in carrying on their race do not seem able to fend off from a considerable portion of their nearer or remoter descendants the same fatal liability to succumb to another bout of unusually severe weather. It is as if the racial scheme of insurance included a permanent provision for such

devastating accidents ; so that the species as such could
rejoice in this guarantee of security, however hardly a rise
or fall in its numbers might bear on the individuals concerned. Between a fluctuating population and a no less
fluctuating complex of circumstances a state of mean
adjustment has been established ; and it is in terms of ratio
rather than of multitude that the chances of survival can
at any moment be measured. Thus the life-force is prudent
enough with all its show of prodigality ; and, though it
scatters abroad its spores and its sperms to no purpose or
profit with all the abandon of a multi-millionaire, it has a
nest-egg laid by. This it does not touch except in times of
need ; when with unimpaired energy it trades on its reserves,
so as to amass a fresh fortune and glut its profligate humour
as before.

But what of Man ? How does all this apply to ourselves, who
have as it were specialized on individuality, and are disposed
to regard it as rather the end than a means of our vital
striving ? For the anthropologist, at all events, with his
extended perspective, the human species is by no means
exempt from biological law, however much inclined its
present representatives in their private capacity may be to
ignore it as a mere bogey. Considering but our last century
of so-called development, anthropological science is bound
to raise the question whether its evolutionary significance
as judged by any permanent strengthening, or simply
altering, of the breed is quite so great as our optimists would
suppose. We may calculate roughly that within the last
hundred years mankind has doubled its numbers. But
such breeding storms are not uncommon in the wider world
of organic life ; and, as we have seen, are wont to leave
behind them no palpable alteration in the quality of the
breeding-stock, so elastic as that is within determinate
limits. Now we can reckon back uncertainly over a few
thousand generations at most in the record of the human
species, even as defined with the greatest latitude, and it is
difficult to read into them more than what are biologically
but minor deviations from a pretty constant mean. True,

the further back we go—having to trust more and more to a somatology destitute of a good half of its standards, such as for instance those "race-marks," as they are loosely termed, which consist in the texture of the hair or the colour of the skin—we seem to detect anatomical differences which tempt us to divide the Hominids into species and even genera. So, too, there survive among modern men divergent types, the white, yellow and black, to follow the popular nomenclature, which in origin doubtless go back to what Bagehot has called the "race-making" period; when it may be presumed that the environmental control was more severe, inasmuch as less culture spelt more risk. As it is now, however, none but these latter variations are preserved; and these are merely intra-specific, showing no signs of being able to interfere with cross-breeding in its germinal aspect, whatever sexual selection may do to curtail the actual process. As with the so-called "local races" of birds, then, we have to recognize what is rather a *de facto* than a *de jure* departmentalization of modern man's breeding efforts; and it is quite possible, nay, probable, that the *homme moyen sensuel* is distributed quite impartially among the existing population of the globe. This generalized man, however, it must be remembered, is in one respect a fluctuating quantity, however firmly fixed he may be in another. By a selection that may be classed as artificial rather than natural—and sexual selection will have to be brought under the former category—we can, always within the set bounds assigned by the common heredity, confer a temporary, and on a long view non-evolutionary, predominance on one or other of the extremer forms that fall within the margin of fluctuation. Thus we can, by intensive selection of extreme traits, breed small Jerseys or big Herefords; though, left to themselves in a neutral environment, each stock will revert soon enough— in quite a few generations in fact—to a middle size. So, again, our bachelors and spinsters, whichever of them do most of the choosing in the matrimonial market—and, for myself, I would give the benefit of the doubt to the spinsters— can throw the weight of their private predilections on one side or the other in the eternal rivalry between long and short, or dark and fair. Nevertheless, the generalized human being who lives on from one generation to another can well afford to indulge in all these concessions to fashion, since many changes of raiment are always to be found in the germinal wardrobe. Nay, a point may be reached when artificial and natural selection come to be completely out of touch, and the former, instead of being merely non-evolutionary in its relation to the race-progress, ventures to become anti-evolutionary ; whereupon disaster follows. Thus the Great Irish Elk perished, together with his wife, because she would have him wear fantastic antlers—a fact which the supporters of female suffrage must explain away as best they can.

Now this apparent impermeability of the germ-plasm to our well-meaning suggestions that it should conform to our ways rather than to its own may prove rather disappointing to those of us who imagine ourselves to have the welfare of the species at heart. But are we sure that the object of these benevolent aspirations is really the species as such—not mankind as it exists now and is likely to exist in the calculable future, but just *Homo sapiens*, with a germ-plasm that may or may not succeed in evolving into *Homo sapientior* ? When the geologists speak of the possibility of another climatic oscillation—one that for instance might compel all Scots to desert Scotland and invade the South like an exodus of lemmings—can any good come of speculating what kind of man is going to come out on the other side of the squeeze, always assuming that there will be a further side to it when " Home to the Highlands " once more becomes a practical question ? Let us at any rate reserve preoccupations with such far-off unhappy things for our more speculative hours. As sociologists, I venture to contend, we are not concerned with them at all. It is enough for our special purposes that we should consider what is being done, and what can be done, to change the conditions of the social life so far as it touches ourselves. This means that we must focus our attention, primarily, on what I have called the social present, a matter of about a hundred years, and, in a secondary way and less intensively, on the moderately spacious domain of cultural history, fading off as it does into prehistory as soon as a few millennia have been traversed. Even if, racially speaking, the game of life has throughout been played by Man with one and the same pack of cards, there have been shufflings innumerable which, from the social point of view, offer combinations on which observation and analysis can be most fruitfully expended. After all, these shifting conditions make the game for the players, who happen in this case to be ourselves. We play in partnerships, to be sure, and to a corresponding extent our interest in the game is collective rather than individual. Yet even so, though in our more exalted moments we resolve to live for Society with some fair idea of what we mean by it, I doubt much whether to any intelligible purpose one could try to live for the Race. For I suspect it to be clean beyond the powers of the Eugenic Society to supply the human stock with a new mutation, or even to imagine whether it would take us nearer to the angels or to the apes.

If, then, I am at all right in my biology—and I certainly cannot claim to speak as an expert—it is only by an unpardonable exaggeration that social questions as they affect those heterogeneous bodies which we know as " nations " are alleged to have a racial import. As if it lay between one more or less well-particularized group of folk and another to take charge of the germinal destiny of all humanity for all time ! I do not mean to say that no genetic interest attaches to those types and sub-types which the anthropologist labels " ethnic," without giving the term too precise a connotation, and in any case intending it to mean something that it requires a lot of averaging to translate into fact. These, however, show no signs of being true races in the making, but on the face of them seem rather like physical adaptations that go closely with special cultures and through these with special habitats. Indeed, in such a correspondence between physique and culture as ethnic likeness implies, it is always a little hard to say which condition is cause and

which effect, and that though we keep Lamarckian considerations at arm's length. The nearest biological analogy,
perhaps, is afforded, not by any differences that can be
brought under the head of species, but rather by those
artificial varieties which the so-called " fancier " of dogs or
pigeons—and rightly does he go by such a name—produces
according to his whim from an ancestral strain which under
natural conditions permits of no such extravagance.
Turned out of kennel and cote, and obliged to compete
with their untamed congeners, our bull-dogs and Pekinese,
our pouters and fantails, would soon go the way of all
abortive flesh. These domesticated animals are what
they are, and can remain so, only because they are dependent
on a culture ; and this holds none the less because they are
parasitic on that culture, in so far as in their humble way
they do not help to maintain it. So, too, then, our ethnic
types are so closely associated with states of culture, high
or low, that the real *Naturmensch*—the authentic animal-man
from whom we are all sprung, who lived literally from hand
to mouth without helping the process out with an eolith—
has disappeared as utterly as the wild camel ; so that it is
probably, though not quite certainly, beyond the capacity
of what were once their respective germ-plasms to revive
either of these effete types.

It would seem, then, that, with its protracted childhood,
our race has fatally committed itself to the higher education.
Hence even the generalized man, that common denominator
who theoretically integrates the vulgar fractions into which
the human race has become divided, must be conceived as
something more than Kipling's Mowgli, namely, as a being
permanently specialized for a life of culture and sociality—
in fact, as Aristotle's political animal, with no help for it
left him but to develop along those lines. On the other
hand, this does not mean that most of us may not have
become too civilized to last out when the next geological
cataclysm occurs—in short whenever a Nature which we
can never hope to control completely insists on a minimum
standard of comfort as the price of racial perpetuation.

Like the "fancy" breeds, we may not be suited to rough-and-tumble conditions. Consequently, in falling back on what might be called the central qualities of the stock—those that make for sheer endurance—humanity might well have to dispense with a good many of those types which, as things are for the moment, cut such a fine figure in the show-ring.

Pursuing the same line of enquiry a little further, I would like to ask whether a positive eugenics is likely to succeed in really improving the race, as apart from creating more ethnic diversity than was ever before. For a negative eugenics, indeed, I must confess, I entertain a sneaking fondness. I cannot see what England gains by keeping half a million certifiable defectives on its list of marriageables. Their total number is possibly about equal to that of the rather less definitely certifiable persons of exceptional ability, to whom the psychologists would accord a fully adult "mental age"—the "high-brows," as their intellectual inferiors would call them. Meanwhile the average citizen, in a proportion of about ninety-nine per cent. as against either of these classes, occupies the intermediate degrees in the scale; and it would appear that John Bull, the generalized Englishman, who even so may be something rather different from, and more intelligent than, the generalized man representative of the race, is neither very stupid nor very wise. Since, however, the ethnic type is not likely to better itself by encouraging its defectives to breed after their copious and indiscriminate manner, a little pruning of the stock in this respect would be surely a politic measure. As for a positive eugenics, on the other hand, we are met with the fancier's difficulty that he can only breed for "points," and even so does not always achieve his blue rose or whatever he would like to create. Personally, if I had to try out such a method, I should endeavour to breed for intelligence, always supposing that the psychologists could furnish me with some intelligence-test that really worked. After all, the development of the central nervous system is Man's most striking characteristic; wherefore it is in this direction that breeding is likely to exploit the potentialities of the stock. Possibly conspicuous results might be obtained ; for instance the numerical relations of Aryan to Semite in Europe might be greatly altered. Possibly, again, nothing would come of it. My immediate point, however, is simply to insist that this would be essentially a sociological rather than an anthropological experiment, seeing that cultural considerations would be paramount in seeking our prize-animal in this particular direction. On the other hand, no one, I think, would want to breed human beings simply in order to explore the further possibilities of the race as such. At most it might be some whole-hearted geneticist who, by means of methods for which he would perhaps have some difficulty in obtaining a licence, should endeavour to treat Man as he now treats Drosophila—not having hitherto, as far as I know, benefited the said fly in any way that it would be likely to appreciate, but simply having converted his laboratory into a museum of monsters.

Meanwhile, in the world as it is to-day a conscious eugenics barely exists ; nor has it had any greater vogue in the past. Consequently we must look elsewhere for the causes—I use the word loosely so as to cover unconscious motives—whereby so striking and widespread a differentiation of the human family into ethnic types has come about. To explain it we must go back to our basic principle that the congenital element is in itself a variable which, though uncommonly stable in its equipoise, is likewise, in virtue of the very indetermination of the life-principle which it embodies, a potentiality of opposite reactions. And what holds of the racial factor considered in itself is no less true of the individual member of the species. He, however, displays a modifiability of his own ; this being as it were a supplementary power of adaptation, acting in a way independently, yet always in alliance with the adaptive capacity that is due to his heredity. These phylogenetic and ontogenetic modes of meeting changed conditions with a change of response are not easily distinguished except in theory—a fact which will largely account for the popularity of the

Lamarckian belief in acquired characters, since the plain man can see for himself that use and disuse are factors in adaptation taken as a whole, whether their effects get through to the germinal system or not. The truth, then, seems to be that the variable belonging to race is the permissive principle ; while the executive principle that takes advantage of the permission thus granted is constituted by the other variable consisting in the experience of the individual. Human stature, for instance, is a racial value which the Patagonian and the Pygmy had leave as it were to exploit each in his own way ; whereupon the other variable, which has no further task than to satisfy the individual will-to-live, takes occasion to decide whether a big man or a small is suited to the particular situation. Any such situation involves not only habitat, the supreme condition for the mere animal, but also the entire set of social customs which with Man, the sole inventor of citizenship, provide the efficient no less than the final cause of his private and corporate existence.

Under the term Society, then, as contrasted with that of Race, I mean to include the whole traditional outfit of Man, that essentially social being, and would assign to it, taken in this comprehensive sense, not indeed the entire responsibility for those differences which I have called ethnic, but at least the proximate agency, the finishing touch, that lifts the adaptive change out of potentiality into full being. Be it remembered, too, that the individual as such is adaptable in a full sense that can never apply to the race, with its relatively static nature. No doubt, the further we go back in history, the feebler becomes the influence of culture ; so that, of acquired habits, preferential mating is left in the far-off beginning as almost the sole instrument on the ontogenetic side for bringing out one available pattern of humanity rather than another. Again, in early times there is less mobility ; whereupon each people has to depend on local resources which may fail to include a sufficiency of those mineral substances, calcium, iodine, and so forth, which the bio-chemist believes to be so necessary to the full development of the human organism, both in order that the ductless glands may do their work, and for a host of other reasons as well. I say nothing here of the vitamins, since, as far as I know, they are less unequally distributed. Up to now, however, large parts of the world continue to suffer from these mineral deficiencies, and are doing little to remedy them ; so that, even supposing the phylogeny as such not to be thereby affected, we might expect the population to display continuously the same physique as a result of repeated individual experiences in the way of deprivation. At the same time Nature, always fertile in compensations, might allow them to elaborate each in turn some defensive mechanism, say, a lighter skeleton that nevertheless fulfilled its function, or an immunity from certain epidemic diseases. In the long run, however, as civilization spreads, education and, perhaps especially, knowledge of the sciences will assuredly tend to even up physical conditions, thus leaving humanity more free to tackle the harder task—one hardly begun, some might say—of regulating equably the moral conditions. Whenever that consummation is reached, I should expect our present ethnic types to become less distinguishable, or at any rate less important from any point of view, including that of marriage. Then truly might one become entitled to speak of Society rather than of societies in the plural. Meanwhile it is in the catholic spirit of science that I would have the sociologist do what he can to help that accommodating bundle of potentialities, the generalized man, in whom we all participate, to realize himself, not some day, not after the next Glacial Epoch, but rather here and now in the extended social present, which includes both yesterday and to-morrow. In the words of W. K. Clifford : "Let us take hands and help, for to-day we are alive together."

PART II.

PRE-THEOLOGICAL RELIGION
IN GENERAL

4. RELIGIOUS FEELING

Let me begin by trying to justify my title—Pre-theological Religion. Now theology literally means " god-story "—in other words, any historical or explanatory account concerning gods, or a god. Thus I take it that, if no god is known, there can be no story about him. The question thereupon arises ; When there is neither god nor story, can there nevertheless be some kind of religion ? One might indeed say " godless " in place of " pre-theological," were it not that the former word has an ugly sound, and might therefore prejudice fair discussion of what I venture to propound here as a problem of some scientific interest. For the principle that everything must have a beginning does not seem to apply to any of the major institutions of mankind—to family, tribe and state, to government and law, to morality and fine art, and finally, and above all, to religion. In every case alike, if we work backwards from the present, traces of them persist until they fade out together precisely at the point at which Man himself fades out also. Not one of them can be said to begin at some ascertainable moment covered by the historical record. On the other hand, particular peoples and institutional groups will always be inclined to date their little day from some given sunrise, regardless of the fact that sunrises are never synchronous for mankind in general ; not to mention a second fact, namely, that a fainter illumination invariably heralds the arrival of the full dawn.

Another matter connected with my title stands in need of justification, or, failing that, at any rate of apology. The use of the " pre " in pre-theological may, not without reason, be denounced as ambiguous. For it might imply either priority in time or priority in type ; and it is notorious that anthropologists have a way of treating these two distinct points of view as if for their purposes they were roughly equivalent. Thus some French writers actually speak of " ancient prehistory " and " modern prehistory " on the assumption that it is legitimate to equate the cave-men of early Europe with sundry living members of the British Empire who happen to prefer nudity to the Man-chester goods which our missionaries press on them together with other luxuries incidental to the higher culture. Thus, whereas the chronological savage has the advantage over us—if it is an advantage to have died in youth—by thousands of years, the typological savage is just as old as ourselves, and simply owns a complexion that does not show it. Meanwhile, we anthropologists—for I am just as bad as the rest—are wont to construct what we are pleased to call an evolutionary scale in which either kind of primitiveness is drawn on indifferently to supply the required gradations, short of the appearance of civilization which becomes the affair of another set of experts. In so doing, then, we are admittedly providing the civilized man with a sort of synthetic ancestor in the shape of authentic bones from Mentone or the Dordogne clothed with the flesh of Australian aborigines, Tasmanians, Bushmen and so forth. Are we therefore to suppose that such a composite picture of human origins can be true, even in broad outline ? For myself

I am prepared to hold that, however unsatisfactory, it is the best that can be done on the existing evidence ; for it is quite impracticable to carry back the actual history of any modern savages past the seven thousand years or so that would give whatever still remains to them from such far-off beginnings an absolute priority over civilization in any of the known forms. Thus the anthropologist finds himself no better and no worse off than any other naturalist who arranges his plants or butterflies or birds in groups primarily determined in each case by the sum of internal resemblances, but thereupon connected one with the other in what is fondly hoped to be a genetic series by reference to certain common features that may be supposed to vary according to the degree of hereditary affinity. My first task, then, will be to assemble the characters constituting systems of belief and practice which, although lacking any definite idea or theory of a god, are yet in other respects sufficiently like the religions that include the notion of a god to be worth provisionally classing and comparing with them. My ulterior object, however, will be to suggest that this simpler type is also earlier ; in short, that religion goes back to the twilight stage of mind and society, whereas a theology implies that the sun is above the horizon and perceptibly mounting behind the mists of a dubious yet fairly hopeful morning.

Now, having much to say and very little space in which to say it, I must refrain from controversy as far as I can. It is only fair, however, to point out at the start that my views are likely to run counter to those of acknowledged authorities in the same field. Of these let me mention but two who in their several ways might be expected to declare that religion and theology are correlative and contemporaneous. The first is Father Wilhelm Schmidt who, if I understand him right, postulates an original theism, presumably revealed, which by a process of degeneration has become obscured by various superstitions so as to lose its hold altogether on the larger portion of uncivilized mankind. He takes his cue from Andrew Lang who, in *The Making of Religion*,

83

published in 1898, produced a very miscellaneous list of what he described as "high gods of low races," whose genesis could not in his opinion be explained in terms of the Tylorian animism, at that time popularly regarded as a "key" to all mythologies. Meanwhile, Father Schmidt in supporting Lang unsheathes a fresh weapon in the shape of a diffusionist method which seeks to establish for the primitive world, taken region by region, a sort of stratigraphy whereby the cultural layers left by successive waves of influence can be distinguished in the order of their deposition. Such a method, all must agree, would be admirable could it be made to work. But can it? For one thing, no confirmatory evidence from the side of archæology is forthcoming to prove any constituent part of existing savagery to be really old as contrasted with old-fashioned in appearance. Again, the method tends to identify the oldest cultures with the most submerged; so that, if theism were part of a primal inheritance nowadays retained only by Pygmies and the like, it would seem that its survival-value was indeed low. In any case, I should be loth to assume that the spontaneous element in any live form of religion was less characteristic than the derived. But, if this factor be equally important, then the constant process of assimilation and redintegration to which successive borrowings would be subjected must effectually disguise the precise shape in which a given contribution was either offered or accepted. Lastly, quite apart from the question of method, I prefer to stand or fall with the biological sciences in treating degeneration as at most a secondary aspect of what has on the whole been an evolution and emergence of the human spirit from lower forms of mental life. At the same time I freely admit that, on the showing of Lang or of Father Schmidt, room must be found for gods of a kind surprisingly far down any evolutionary ladder that the facts allow one to construct.

The other authority with whom I expect to find myself in disagreement is Sir James Frazer, who would roundly stigmatize as magical the whole class of godless rites which

84

I would assert to be religious in their underlying motive. It is more than a question of words between us, because Sir James Frazer distinctly attributes to them a separate intention, namely, that of exercising control over Nature in a mechanical, though as it happens a pseudo-mechanical, way. Thus when religion provides the idea of a god, that is, a personality superior to Man, who must be conciliated if he is to use his power over Nature on Man's behalf, a downright revolution of attitude must be supposed to have occurred, a sudden shift over from a mechanical or would-be mechanical, to an inter-personal, relation. How and why such a change comes about is not made very clear ; but the bare suggestion is thrown out that the failure of magic was religion's opportunity. Apparently sheer experiment using self-interest as its only standard persuaded mankind that religion gave better results ; though it might be thought that experiment had sometimes seemed to prove the opposite, since it is notorious that when religions grow obsolete—as happened, for example, with the paganism of the Roman world—they tend to resume what on all sides would be described as a magical form. The root of my objection, then, to the Frazerian position is that it implies an interruption of continuity, where I would rather assume an unbroken development. As for magic, if the word is to retain that unfavourable connotation which it is both usual and convenient to attach thereto, it seems to me better to suppose that, together with religion, it is differentiated by antithesis out of much the same stuff, namely, an experience of the uncanny giving rise to confused and often conflicting valuations of the power therein manifested in its bearing on human welfare. But all this will be made clearer presently. For the moment, it is enough to indicate that I am prepared to find plenty of nascent religion in that mass of material, collected in *The Golden Bough*, which its indefatigable author labels " magical," and yet, as it were despite himself, offers as a key to the so-called religious phenomena ranged side by side with them in the same treatise.

Without further prelude, then, let me proceed to consider

the beginnings of religion, seeking for them first and foremost
in the domain of feeling. For working purposes the conception furnished by Analytic Psychology of mind as a trinity
of functions, constituted by feeling, thinking and acting,
will serve us well enough. Moreover, Genetic Psychology
treats feeling as the basic element that finds expression in
acting with an intermediation on the part of thinking that
does not become effective until mental evolution is well
advanced. The chances are, then, psychologically speaking,
that if religion corresponds to any ingrained disposition of
human nature it will come out in feeling, and through
feeling in acting, before thinking can get to work on it.
Just as the child wants and tries to talk long before he can
express himself articulately, so, I believe, it is with religious
experience ; which needed and sought a god, if one likes to
put it in that way, long before it knew him by name. So
let us consider the genesis of religion first of all in the light
of the impulse or set of impulses that it arose to satisfy.
Since religion is ultimately an affair of the mind or soul, we
may be pretty sure that, however we may fare later on, this
is at any rate the right way to start.

In our reconstruction of the mental history of *Homo
religiosus* we can surely imagine him as already a stage or
two ahead of any other known animal, the anthropoids
included. We can, for instance, assume him to be already
an omnivore, and as such to have a diversity of regions and
climates open to him. We may also grant him fire, since
Sinanthropus, not to speak of the Neanderthals, had it ;
and with fire would come into existence the fire-circle and
the makings of a home. Further, let us give him a rudimentary language, no doubt eked out with much gesticulation.
As for the state of his arts—his use of weapons of wood,
stone, and bone, for instance, which goes back a long way in
more or less calculable time—we need not furnish him too
lavishly with such artillery, since weakness in that respect
must be counterbalanced by numbers, if he was to stand up
to big and dangerous animals with any chance of success ;
and that he actually did so we know by the bone-middens

86

that he has left behind him up and down the strenuous wilderness of a glaciated Europe. For numbers have made Man what he is. Because he hunted in packs, he had to devise for himself a pack-law, as Kipling calls it ; and the political animal in him—or rather in them—was launched on its career.

This fact is all-important for our present enquiry, because the political animal and the religious animal have a great deal in common. With us no doubt political organization and religious organization tend to go their several ways with results disastrous to both ; but, for the savage, his state and his church are one, and his conduct is shaped under the joint influences of a secular and a sacred sanction. How this should come about is really not very hard to understand, at any rate in principle. But, before completing our sketch of the bare essentials of that early condition of mankind coeval with the birth of religion, one more touch must be added to our imaginary portrait of the primæval community. Not only must it be conceived as fairly numerous, occupying a cluster of rock-shelters or a roomy cave at the least ; but it must likewise be credited with the rudiments of a polity, in a system of rights and duties based on kinship—such kinship being almost certainly reckoned through the mother rather than the relatively irresponsible father, whenever consciousness of descent came to the surface of the mind. Nay, being such a natal association, it was a true community all along, having something more elemental and warmer than mere gregariousness, namely mother-love, to provide the very framework of the loom on which the criss-cross of human relationships was to be woven into an elastic but lasting texture.

Now, if one had to sum up in a word the specifically religious feelings, forming as they do a very heterogeneous compound or complex, one might perhaps use " awe " as most nearly presenting the required blend of meanings. Awe—which unfortunately does not easily lend itself to translation into other tongues—is not equivalent to pure fear, but stands rather for a submissiveness tempered with admiration, hopefulness, and even love. Thus on the whole it betokens an attitude, not of bare avoidance, but on the contrary of shy approach. And, next, what of the object towards which awe is felt ? Generalizing somewhat boldly, let us term it the uncanny. Human life is ever beset by uncertainties, and not least of all at the gathering and hunting stage of society, when quite literally one's next supper depends on human providence *plus* a fair slice of luck. It is this incalculable complement to Man's ordinary ways and means of maintaining himself in being that sooner or later provokes attention on its own account. Once aware of a mysterious control operating on the fringe of his vital efforts so as sometimes to further and sometimes to thwart them, he is bound to pay an uneasy respect to whatever it is that thus over-reaches and over-rules his natural powers. For luck, good or bad, is more than chance, being chance as bearing intimately on our personal interests and thereby brought into a relation with us that seems to impose a purposive character on it despite itself. It ceases to be the arbitrary behaviour of an indifferent world-process, and becomes the way in which my universe happens, and, one might almost say, chooses, to behave to me. In leaning upon his luck, Man charges it with a certain responsibility, so that the unknown remainder which is so much more potent than what his little knowledge can command is henceforth regarded as if it had a diffused will of its own. My universe in short is ultimately like me so far as to permit a certain correspondence between our aims ; and it might even be felt on the analogy of human intercourse that good intentions on my part were sure to be reciprocated—in other words, that they might be expected to turn the luck in my favour. As for bad luck, since " conscience doth make cowards of us all," it might well be taken for a sort of punishment not inconsistent with a general readiness to befriend the righteous ; the latter for their part doubtless taking as much credit as they dare for such good fortune as befalls them. In such nebulous intimations of help from the Beyond, indeed, we may perceive the faint adumbration of a God, or at all events of a Providence ; for in any true evolution the end must be implicit in the beginning. But such an attribution of an interest in Man to a power outside Man's ken is not only too vague but, as we shall now go on to note, too occasional to be classified as more than potentially theological—that is, pre-theological, as I would have my phrase understood.

For primitive intelligence is subject to two chief drawbacks— obscurity and intermittence. Just so when the ancient cave-man, bent on outfacing the dark and its secrets, took his stone-lamp in his hand—the specimen from La Mouthe with its traces of burnt fat in its shallow basin tells us exactly how he went about it—not only was the light dim, but, with all the care in the world, it must have been exceedingly liable, if not to go out altogether, at least to flicker. Thus it is only by glimpses, as it were, that the religious experience of the early Man makes acquaintance with the environing forces that seem to be peculiarly concerned, and hence to concern themselves, with his private fate. Recognizing this, anthropologists work backward from polytheism to an even more incoherent polydaemonism. Let me, then, offer them a polydynamism to complete the regress. Usener, indeed, has already provided such a *terminus a quo* in his *Augenblickgott*, or " momentary god " ; but with all respect I deem such a term as " god " or even " godhead " an abuse of scientific terminology, which cannot indulge in *prolepsis* unless anticipations and attributions are to be taken as equally constitutive of the same fact. For awe, at any rate until it is supported and explicated by an organized cult, is excited by the uncanny, not least of all for the very reason that it comes in the guise of a sudden flash out of the darkness. Its sheer accidentality throws a mind inured to a humdrum routine off its balance, so that it does not know what to expect next, and yields to a sense of bafflement more or less pregnant with such hope as may counteract the anxiety. This is the point at which the question " Is it for good luck or for bad ? " is bound to obtrude itself. The situation has taken and shaken the man out of his lethargy, and yet, though inwardly active, he must remain outwardly passive, so long as he can find no certain line to follow. He has the nightmare feeling of trying to run, and yet of being held up. No wonder that he tends to externalize that which urges him ; seeing that the other part of him, which after all represents his normal nature, finds itself impotent to make any effective response. Gradually, no doubt, he and his fellows will contrive some technique for dealing with what might be termed crisis in general. For the moment, however, we are supposing the individual or the group—and we may be pretty sure that at first it will be group-experiences that mostly count—not to have at disposal any ready-made procedure for dealing with a particular visitation ; so that we may the better perceive the workings of a mind so taken aback.

Much attention has been paid in recent times to repressions and their cure, and, although a lot of rubbish is both preached and practised in connexion with the subject, there is also some sound psychology at the back of the movement. Thus it is certain that pent-up emotion will out somehow, if sufficiently violent. Again, it is pretty clear that, when the normal channel of discharge is blocked, so that what may be termed the primary activity is out of the question, the secondary or substituted activity whereby relief is gained will have to imitate an effective reaction closely enough to afford the feelings an excuse for letting themselves go off on a side-track. Symbolism, in a word, is in its essential function a spiritual lightning-conductor. Even though ghosts be bullet-proof, a sufficient expenditure of blank cartridge will probably terminate the interview in favour of the shooter—at least, so I should expect without having tried it. Thus it does some primitive community stricken with small-pox a world of good to carry out an image of the disease-demon and douse it in the nearest stream. Nay, when Neanderthal man furnished his dead with all the materials necessary for a post-mortem existence, was he not dimly aware that it was but a *pis-aller*—a " worse-go " that was nevertheless a lot better than no go at all ? For I do not believe for a moment, that, as a certain learned Professor once suggested to me, *Homo primigenius* was simply so stupid that he could not recognize death when he saw it, but carried on as usual until it slowly dawned on him that he was wasting his time. On the contrary, I am convinced that death—not perhaps any death, since slaying animals, and it may be his enemies as well, was part of the day's work, but at any rate the death of his nearest and dearest—was a catastrophe, an event entirely subverting the due order of things, and therefore one that correspondingly found him at the end of his accustomed resources. Yet what an assuagement of the horrified consternation of the bereaved cavehold to make elaborate pretence of furnishing the wherewithal for a continued inhabitation, or a journey, or whatever active proceeding the poor corpse was manifestly incapable of carrying out in its fast-decaying flesh. It needed, in fact, the additional counterfeit of an invisible something, that utilized the visible grave-plenishings in its own invisible fashion, to make any sense of their actions ; if sense, of the practical materialistic order, were even demanded so long as their mood of mourning lasted. Now our archæologists in denying the appellation of *sapiens* to this lowly race would almost seem to deprive them while yet alive of any soul worth mentioning. Yet I strongly suspect that they had so much of it as to spare a little for their dead ; not, I daresay, *totidem verbis*, since their verbal experiments may have been strictly limited, but by a self-projection none the less satisfying because its vehicle was but a symbolic action, that is to say, an unreal or pretended doing with a real meaning.

We may next pass on to enquire how out of a diversity of odd, unco-ordinated reactions of this secondary or symbolic type, as provoked by one critical situation or another as it might chance to occur, there was gradually formed a religious custom, whereby that which on its own part happened irregularly was nevertheless dealt with by rule from the human end. Now, to begin with, a certain talent for

91

imitative, and hence significant, movement is the common
possession of the human stock. Indeed, I have seen it
stated that our own deaf and dumb can successfully collogue
with savages all over the world by means of this rudimentary
language. At the same time every human group is likely
to produce its dramatic genius, who by natural or, as he may
well claim, supernatural inspiration gives a fresh turn to the
stock expression of those intenser feelings that are especially
contagious. Thereupon the others will gratefully adopt the
innovation, so that in time a full service of " Gestures
Ancient and Modern" becomes available for suitable
occasions. Meanwhile the maintenance of such a ritual
tradition depends almost entirely on the older men, and—
whenever the female sex has ceremonies of its own, as
the sex-solidarity of savage society is sure to entail—on
the older women as well. For it stands to reason that the
crises of each rising generation will in large measure become
the commonplaces of those with a much larger experience
of the ups and downs of life. Age grows seasoned to all
the pranks of fortune, and gradually learns to meet the
unexpected half-way with a callous audacity that of itself
helps to parry its worst blows. Thus by setting up a convention of behaviour such as is at least suggestive of a
prosperous issue, and hence by association comforts and
braces the feelings, these greybeard mystagogues provide
the group with a sort of universal insurance against risks.
Nay, at this point the occasional nature, which at first
belonged to the abnormal event that awakened awe,
would in large part give way to a periodic status accorded
to it in a scheme of rites conforming to the cycle of the
social and economic life in general. Thereupon a certain
centrality of interest is able to develop, the effect of which
will be to invest the various contingencies of existence with
a like degree of common meaning and purpose. Thus they
will come to be referred to a sphere, or rather hemisphere,
of their own. The scheme of rites as a whole will be
directed towards an equally wholesale effect. Let us go
on, then, to consider what aspect this will assume, taking

it for granted that at first it will be something rather indefinite.

We may start by distinguishing as derivative rather than intrinsic the character that will be attributed to the object of religious experience simply in virtue of its close connexion with the rest of the customary round of daily and yearly occupations and amusements. As soon as any ritual becomes periodic, it acquires the respect due to it as part of a hallowed tradition. Such is primitive conservatism that to be old and to be valued almost amount to the same thing. Nay, just because immemorial usage will be least intelligible on its ceremonial side, a blind faith in the wisdom of those that went before will entail a no less passive acquiescence in the most burdensome of such prescriptions as most completely defy explanation. Again, periodic rites taken in conjunction with the ordinary business of life will punctuate the latter, as it were, with intervals of another kind of activity—one less akin to drudgery and more exciting, or at any rate more intriguing, to the spirit. Thus the festal mood will be apt to make common cause with the religious ; and that though religion has sterner work ahead than to wander down the primrose path of the fine arts. Once more, rites will tend to entwine themselves with all manner of specialized tasks such as feeding, fighting, marrying, burying, judging, healing, and so on ; and to the same extent the mystic influences thereby brought into play will become as it were departmentalized, and assigned each its separate place in a sort of functional pantheon. Here, then, are various ways in which there might be imputed to the order of wonder-working things in their entirety a chameleon-like habit of simply reproducing the tint of the social surroundings of the moment. Nay, we shall presently see that, were it not capable of becoming the reflex of our social feelings as well as of our awe pure and simple, it could never become fully moralized at all. Meanwhile, there is reason to think that it is primarily objectified under colours of its own, so to speak. Years ago I proposed what I called "the *tabu-mana* formula" to describe it in what might be distinguished as its existential, in contrast to its moral, aspect. Let us, then, see what characters are involved in this conception.

Now, grammatically speaking, both *tabu* and *mana* would seem to come nearest to what we understand by adjectives, though the liberal usage of the Polynesian languages whence they are derived enables them also to play the part of substantive or verb more or less as required. One might almost venture to say, then, that there is an adjectival as opposed to a substantival stage, or at any rate phase, of primitive religion, when it generalizes its object as it were distributively, taking note simply of a quality, perceptible in all manner of things, of sometimes behaving oddly and in like degree warningly. Any portent as such—what the Greeks call a *teras*—combines these two conditions of being at once strange and ominous. *Omne ignotum pro mirifico.* Hence the plain man—and he is as typically religious as any of his betters—stands towards it in the first instance on the defensive. *Tabu* does not stand for fear, which is emphatically not the basic element in religion, but for caution, which is quite another thing. Prudence enjoins a certain self-inhibition, a sort of crouching attitude that serves the double purpose of enabling a man to lie low and to gather all his forces together. The genuine *tabu* feeling is not paralytic, but pregnant. The hunter pauses in tense expectancy, having an inkling that there is something big behind the bush, whether it mean his dinner or his death. Correspondingly, then, as applied to the object *tabu* signifies, as Codrington well renders it, "not to be lightly approached." It implies a character that is not so much neutral as ambiguous, because it is bound to react forcibly one way or the other. We may say, in fact, that the subject reads his own state of tension into it. *Mana*, which is usually translated "super-normal power," is just this high voltage, a potentiality of opposites in the way of weal or woe, but in either case something beyond the measure of workaday experience. In imputing, then, to the savage a *Weltanschauung* conceived, or rather perceived, in terms of *tabu* and *mana*, one is simply supposing him to be aware that, so far as life does not consist in routine, it is exceedingly risky ; and that, taken all together, the risks call for greater heed because their issue is more momentous. He has to teach himself to run on two gears, the one easy because jog-trot, the other exciting because neck-or-nothing. I do not suppose that he seeks religion, because the human animal in us may well prefer to grow fat in safety and somnolence. Rather religion finds him, since his fairway is beset with many hazards, and, once in them, he must perforce play out bravely or else quit the game.

So far we have been conceiving this wonder-world, this realm of unaccountable happenings, as something unmoral, like the weather or the sea—just an unsteady side of brute Nature. And so it might be, if it failed to respond to Man's efforts to humanize it, as on the whole the weather and the sea have succeeded in doing up to the present. But it is no illusion that we have considerably reduced the risks of living by facing them with the right mixture of discretion and confidence. We have done this partly in a material way, but partly too and perhaps chiefly in a spiritual way. Thus on the material side we have improved our arts by way of our sciences ; so that a lot of the old bogeys have simply vanished, like rats from a house that has been thoroughly cleaned and repaired. This improvement, however, in what might be termed the sanitation of life is for the most part a direct consequence of our progress in mental development. Whether the part played in modern education by religion is as salutary and effective as it ought to be is an open question. Historically speaking, however, there can be no doubt that religion has presided over most of that long process whereby the human mind has attained to a relative freedom from primitive credulity, so far as it takes the form of a childish fear of the dark. By dominating to a notable extent the physical dangers of mundane existence we have given ourselves a chance of concentrating on the spiritual dangers ; which in our overgrown and amorphous societies of to-day are perhaps more insidious than ever.

How, then, has this moralization of the raw material of religion taken place, if we are right in supposing its original concern to have been with good luck and bad as they crop up promiscuously, the mushroom with the toadstool. We have already noted that Man is egocentric enough to deem nothing in Nature irrelevant to his interests. Thus he takes it for granted that somehow there is purpose in it all—that, whether lucky or unlucky, these things are aimed at him. More especially, in so far as they amount to life-and-death matters, does he construe them as deriving their immense significance from their bearing on his private fate. Thus he holds his universe in fee from the first, so far as he treats it all along as subject to his moral judgments. He expects it to answer to his praise and blame, though experience shows him that it will not do so except according to a whim of its own which must be humoured. This business of humouring, then, as exerted towards the scheme of things at large, at first just when it happens to call for such treatment, but later on as part of a customary round of courtesies, becomes the chief preoccupation of the leader of the primitive society. Such a master of ceremonies is never so much needed as when the people are beside themselves either with too much joy or with too much sorrow ; for, besides lacking internal controls, unless their equivalent is provided by some social means, the savage is pretty sure to come to grief under one or the other form of emotional intoxication. In such tumultuous circumstances, then, wise elders invariably fall back on the same prescription, and make a dance of it. From dance to prayer might almost be said to sum up the history of religion regarded as a mode of human self-expression. To dance out one's transports, whether pleasurable or the reverse, is to impose on them, that is on oneself, and by proxy on one's universe, the quality which in turn is the essence of the moral law, namely, measure. Instead of mastering the man, the feelings thereupon acknowledge his mastery, and he—that central selfhood of him, which is the final arbiter of good and bad —is flattered and uplifted. Nay, the feelings themselves

take on a new character, inasmuch as they are no longer
isolated discharges with a sort of whole-or-none reaction
liable to shake the organism to pieces. Instead they become
harmonized, so as to lend severally just as much force as is
required for some symphony of the passions that will be all
the more rich and satisfying for being chastened and self-
possessed. It may be added that such a sobering of crude
sensibility favours the growth of thought, which thus
quietly comes to its own as chief assistant of that ultimate
choir-master, the rational self or will.

Apart from its function as a discipline and purge of the
feelings, the dance—to use the term as a shorthand equivalent
for the ceremonial activities of the primitive world in
general—is typically congregational or communal in character ; and herein we have a second source of its efficacy as
a moral factor. For, whereas any freakish or untoward
event is apt to set the savage dancing as a way of neutralizing
the shock, and as it were spreading the effects over his
system, he dare not dance in public for a shameful reason ;
nor indeed without the co-operation of his friends can his
effort amount to more than a travesty of real dancing,
which is always choric. If, for instance, hate or lust predominate in his fit of violent excitement, he must work off
his repressions in secret, no doubt reproducing in a symbolism of some scurrilous and hole-and-corner type certain
borrowed features of those demonstrations in which the
community in a body permits itself to indulge in reference
to the public enemy. This way, in short, lies black magic,
which, just because it cannot show its face abroad, is unable
to attain to a procedure containing enough inner meaning
to evolve on its own account. Wherefore the sorcerers of
all ages, when stripped of the second-hand rags in which
they trick themselves out, are found to be as dirty and
poverty-stricken in their ways as when they started. On
the other hand, the society as a whole may be assumed to
act for its own seeming good ; and, though the moral
horizon of the savage may be narrow and, so to say, domestic,
he is at least loyal to his group, and supports the common

cause with a faithfulness which we with our more mixed, if more far-reaching, allegiances might envy for its intense quality. Hence he is too good a patriot to be sceptical about the least jot or tittle of his custom, including his religious custom. That it all makes for the general good is axiomatic with him. Thus when he dances he does so with conviction. He meets the remedial function of his ceremonies half-way by believing that the cure cannot fail. And no doubt this is the main secret of his survival amid all the hardships and perils that ever threatened nascent humanity with an untimely ending. Rightly or wrongly—and in view of Man's record we must allow that it was in the main rightly—the savage was confident that he had taken the measure of the chances of this mortal life, when he had somehow made them responsive to a rhythm of his own composing. Of a universe so attuned to his desires he could no longer be afraid, but might even go on to offer it a shy friendship.

Now in claiming some such outlook for primitive religion, one must hasten to add that all the while feeling was well in advance of thought ; so that we must beware of making a philosophy out of what was at most an æsthetic appreciation of the dance as a means of getting into touch with good and evil in their more unlimited forms. The values obtained are simply immanent and self-evident to the feelings. Every crisis tends to be met in the same way, namely, by a ceremonial movement symbolizing first a tension, more or less prolonged, and then a release celebrated in a sort of major key that takes the place of the previous minor. For there is always this progression in the emotional tone—from *tabu* to *mana*, from Lent to Easter. Whatever shape the extraordinary experience may take—whether it come as a great darkness or a great light—it numbs the senses for a while. Hence, pending the recovery of self-mastery—always a joyful consummation—there is a certain depression, carrying with it a temporary discouragement that may readily translate itself into a disgust of self, a conviction of sin. No doubt it is only the finer spirits—those twice-born hiero-

phants whose individual inspirations are collectively en-
shrined in the established mysteries—who are at all acutely
sensitive to this transition *de profundis ad altiora*, this experience of " dying to live." The ordinary savage, on the other
hand, has to be pretty tough in order to hold his own with
a hard lot, and is anything but a neurotic. Yet, as everyone
has to be ground in the same ceremonial mill, he too
cannot help being conscious in dimmer fashion of the common
tendency of all his rites to combine a piacular opening with
a communial, or sacramental, completion. Nay, it is not
fanciful to connect with the difficulty of otherwise making
an impression on the type of mind that goes with a leathery
hide the tortures—as we at least would rate them—that so
often are provided by savage custom as fitting accompaniments of the *tabu* condition. The valley of humiliation, the
temptation in the wilderness—these agonies that would
seem to be a prior condition of the exultations in religious
experience at all stages of its development—must take the
form of literal gashings and scourgings for those who are
as yet susceptible only to concrete symbols, to corporeal
metaphors. Similarly their sense of final release may lead
them not only to feel, but to feed, like giants refreshed.
Nevertheless the coarser imagery must be discounted if we
would discern the spiritual purpose implicit in the rite
that achieves *mana* by way of *tabu*. From the very caprices
of Nature the child of Nature has learnt how to turn bad
luck into good by enduring the fast until the feast is ready.
So much, then, by way of a sketch in roughest outline
of the general tendency of a type of religion which is pre-
theological in the sense that it uses adjectives rather than
nouns to express the appeal of the divine. Nay, one might
expect the very adjectives to be missing—though this happens
to be quite exceptional—because the felt qualities for which
they stand correspond to moods directly associated with
symbolic activities that can serve in themselves as adequate
vehicles for those whose minds work as it were through their
muscles. Neither leaders nor followers have as yet acquired
powers of reflexion such as only a command of words

can bring into play ; but if, when face to face with the
unexpected, they would find it fruitless to pause to think,
they have at least learnt to pause to take a long breath,
recognizing that it promises a stronger pull in the end.
So far as we have gone, then—and it is to be remembered
that up to now we have been viewing the matter from one
angle only—it looks as if the savage had already begun to
moralize his universe by thrusting on it the rôle of a rather
rough-handed teacher. At anxious and decisive moments
he beats a wise retreat into the inner stronghold of his own
character, and there discovers himself to be in better fettle
than he knew, and so all the fitter for the next bout. Thus
he can more than half forgive a plaguey world that has
driven him to fall back on spiritual exercises such as turn
out to have value in themselves quite apart from their
pragmatic consequences. Some primitive folk, such as the
Arunta of Central Australia, make up for a miserable
environment by spending half their time in a transcendental
region where they commune with their ideal selves in the
shape of reincarnating totemic ancestors, and after their
own Stone-Age fashion taste something of life's inner
meaning, though with never a God in sight. Wherefore
with many similar examples in view of dealings with the
unseen on the part of primitive man that, however blindly
directed, have somehow and somewhere found him what he
wanted, namely, a strong heart, I venture to ask whether
even a godless type of religion may not afford a partial
revelation of the Good as it evolves in us and for us.

5. RELIGIOUS THINKING

Much the hardest part of
our triple task is to attempt to do justice to the intellectual
element in primitive religion. It is not that the savage is
deficient in thinking powers of a kind. But he thinks, so to
speak, concretely and not in abstract terms, the very words
being wanting in which such abstractions could be presented
to the mind. We on our part, however, are so well accustomed to rearrange our experiences under general heads,
and to deal with the categories so resulting as if they were
solid facts of a higher order, that we find it hard to avoid
the psychologist's fallacy, as it had been called, of reading
our own type of mental habits into those whose ways are
not as ours, but, whether better or worse, at any rate
different. Psychologically, that is, historically, viewed, the
wilderness is not simply a garden mishandled or run to
waste ; and the lustiness of growth responsible for the
primæval tangle needs to be appreciated on its own account,
however much the accompanying untidiness may offend
suburban taste, or comport ill with the demands of a modern

vegetable market. For the scholar it is especially difficult to throw himself back imaginatively into a world devoid of all book-learning. Let there be no Bible or Koran to serve as a repository of sacred lore, and unsupported memory becomes sole custodian of the oral tradition.

Now we must not make the mistake of underrating the powers of human memory when trained to rely on its internal resources. We civilized men depend on false memories very much as we do on false teeth, and, surrounded with works of reference, hardly note the *caries* to which the native article becomes subject through such soft feeding. Anyone, however, who has been at all intimate with the illiterate peasant will have marvelled at his retentiveness of the minute details of what we might reckon an almost featureless lifetime ; and, though on certain topics he will be inclined to embroider, as any good story-teller has a right to do, he will achieve remarkable accuracy if a given matter, a date or a price, strikes him as worthy of meticulous telling. Transmission from mouth to mouth, on the other hand, is more likely to fail in exactness, unless disciplined by conscious art as among those who make it their profession to keep such folk-memory alive. Indeed, what may be distinguished as the exoteric part of primitive tradition is frankly imaginative in tone, and it becomes a chief function of the past to enwrap the present with a luminous atmosphere of wonder. *Il n'est miracle que de vieux saints*, says the old French proverb ; and in the presence of antiquity the unsophisticated are always childlike enough to beg for a nursery tale. Whether simply meant to amuse or given an edifying turn, such efforts to picture fact as it might be rather than it is belong to the recreative side of the social life, and must in fact be classed as a kind of play of the mind. Religion, on the other hand, though in virtue of its recreative function it is generically akin to play in the wide sense that makes the latter co-extensive with all activity pursued for its own sake, is at the same time severely practical, in that it would take charge of life as a whole, directing all its activeness towards the realization of one all-pervading and all-satisfying purpose.

Now, historically speaking, religion may be hailed as the mother both of the sciences and of the fine arts, including literature. But the two broods come of different fathers, Common Sense in the one case and Play of Fancy in the other; and we may fairly say that, the more they grow up, the more does the paternal strain come out in the children. Thus, whereas what is usually classed as mythology when we do not ourselves happen to believe it to be true covers a mass of apparent statements of fact concerning a miscellaneous personnel of gods and devils, heroes and figures of romance or sheer fun, its connexion with religion in the sense of cult is often so slight as to be virtually non-existent. One might indeed go so far as to say that religious doctrine in our sense is to be found only in one class or rather sub-class of myths, namely, that special kind of ætiological or explanatory story which purports to account for a given piece of ritual—the mystological or hierophantic type, if we may label it thus. Even so, we must have evidence that it is actually organic to the rite in question—that, in fact, it somehow forms an oral accompaniment of the ceremonial scheme by providing substance or ground or at least colour for formula, prayer, hymn, or revelation—before any importance can be attributed to it, apart from its intrinsic interest as a wonder-tale. To take a casual example, we hear that Demeter, sorrowing for the loss of Persephone, found her way to Eleusis, and that here her spirits were raised by the jests—rather broad jests, one would infer— that were cracked for her benefit by the handmaiden Iambe. Now scholars are probably right in suspecting this to be an attempt to justify those scurrilities which it was proper to utter in the course of the Eleusinian festival, by way of a purification taking the form of a spiritual evacuation—a sort of confession of sins conducted on exhibitionist lines. It will be noted, however, that the Greek apologia entirely misses the real point, when it stresses no more than the funny side of a cathartic process from which the serious-minded must surely have been able to derive a certain moral gain. We may judge this from the fact that in Ashanti, where similar ebullitions of the carnival spirit occur, the priests are able to offer good reason for them on the ground that they render the hearts of the people " cool "—in other words, enable them to get rid of their repressions, as it is now the fashion to say. Another point to observe about the tale from Eleusis is that it almost certainly belongs to the exoteric tradition, since no initiate would ever venture to publish to the world at large the secrets constituting the real heart of his mystery. Such an esotericism invariably defends itself by answering the foolish according to the measure of their folly. Even aboriginal Australia has elaborated this device so as to satisfy its chattering womankind or the too curious stranger ; and it is to be feared that, just as Herodotus on tour in Egypt may have found the temple guides more communicative than the priests—and indeed he himself would have thought twice before publishing what the latter imparted to him in earnest—so many of our anthropologists, note-book in hand, are apt to fill them with fables expressly designed to sidetrack the inquisitive outsider.

It almost comes to this, then, that, before anything in the way of a sacred story of explanation can be treated by us as a genuine contribution to primitive theology, we must require a guarantee of the civilized observer who reports it that the natives had made him free of their rites as if one of themselves. All too rarely, however, for the convenience of our science is such a privilege accorded, and we must gratefully make the most of whatever scraps of inside knowledge are forthcoming from our authorities. Thus both Howitt and the confederate Spencer and Gillen assure us that they passed with the Yuin and the Arunta severally as fully qualified initiates ; though we may doubt if any of them was equal to the appalling task of penetrating to the heart of an Australian language. Captain Rattray, again, an accomplished linguist fully able to cope with the far more articulate niceties of the Ashanti idiom, can boast that, alone of white men, he has been permitted to enter that Holy of Holies, the national stool-house, and has in every way been accorded the local rank of Doctor of Divinity. On

the basis of such authentic information coming from areas
not only geographically distinct but representing about as
extreme a contrast in development as occurs within the
limits of what can fairly be classed as primitive culture,
one might institute an instructive comparison in order to
show the relative importance of the oral element in these
different cases. Thus it would be niggardly to refuse to
credit Ashanti with a theology, even if the attempt has
scarcely been made to reduce it to consistency and system.
Certain it is that no wholesale description of its general
tenour conveyed by means of disparaging terms such as
" fetish " or " juju " will serve to express the degree of
spiritual refinement to which the religion of Ashanti attains,
if viewed, so to speak, from the standpoint, not of its outer
court, but rather of its inner shrine.

Thus Nyame, the ruler of Heaven, is not one of those
so-called otiose deities of whom frequent examples are cited
from Africa ; it being reported that they reign without
governing, like any *roi fainéant*, lesser powers usurping their
active functions so as to attract to themselves the entire
attention of obsequious humanity. On the other hand, it
was only as a result of his privileged position that Captain
Rattray after much enquiry became aware of the fact that
Nyame had a temple of his own ; where, by the way, no
suman or material emblems of the type labelled " fetishes "
in our museums are tolerated at all. Intermediate in status,
that is, superior to the *suman* which appeal chiefly to the
vulgar and gross-minded, but below the High God, though
having no very definite relations with him, are the *obosom*, or
godlings, who in turn seem to differ, chiefly in beneficence,
from the *sasa*, or spirits, including those of the more powerful
or dangerous animals. Applied to man, the animistic
philosophy loses itself among ill-reconciled distinctions.
Thus the principle of life is the *okra*, but the spiritual element
involving character and worldly success—the " personality,"
as one might say—is the *sunsum*. The latter seems allied to
the *ntoro*, which, however, is the undying but reincarnating
part of a man which is inherited in the paternal line, unlike

the *mogya*, or blood, which is transmitted through the mother, and forms the basis of the *abusua* or matrilineal clan. Finally, though the *ntoro* remains on earth when a man dies, his *saman*, or ghost, repairs to *samando*, the spirit-world. Here, then, we have all the ingredients of a theology needing but the baking-power of logic to make the cake set.

Turning now to Australia we had better first consider the Arunta of Spencer and Gillen because, if we go by their vocabulary, they are immeasurably poorer than the Ashanti in conceptions by means of which they can submit their religious experiences to the control of reflexion. Thus in vain do we look for anything corresponding to *Nyame* or the *obosom*—that is, to a Supreme God or even to a host of lesser godlings. Only the *sasa*, those freakish and often malicious spirits which often show animal affinities, have their counterpart in the *iruntarinia*, beings somewhat on a par with our fairies, who are apt to plague mankind with their tricks ; though it is given out by medicine-men, not impossibly with tongue in cheek, that they play a part in the secret initiation undergone by the heads of the profession. But undoubtedly the religious interest of the Arunta centres in the *churinga*, or bull-roarers of wood or stone, which in the language of Ashanti would undoubtedly have to be classed as *suman*—those material symbols for which the High God Nyame has nothing but contempt. Now *churinga* means literally " secret," and it is because the bull-roarer is the central mystery of the initiation rites that it becomes the chief embodiment of whatever the tribe holds most sacred. It serves this purpose because it is something that can be handled—that communicates its immanent grace by physical contact instead of by word of mouth. No doubt aboriginal thought struggles hard to distinguish between the grace itself and its vehicle, but can get no further than to say that an inspired man is " full of *churinga*." Just so another tribe, the Kabi of Queensland, are reported by Mathew to declare of the doctor that his whole inside consists of the crystals that he uses in his cures, so that " hand, bones, calves, head, nails " are made up of them, and he is thereby rendered *manngur manngur*, " alive " in a superlative degree. Plainly the all-pervasiveness, not to speak of the contagiousness, of this vital force is fully realized by the native mind, though the imagery retains its appeal to touch and sight. Thus while fetishism, or whatever such dependence on a materialistic type of symbolism is to be called, betrays itself openly on the surface of such faltering thought, a sympathetic interpretation cannot fail to discern an underlying sense of the difference between the spiritual reality and the outward form of its manifestation. As for his psychological account of Man, the Arunta cannot indeed rival the Ashanti in respect to the sheer multitude of his categories, but, sharing as he does the belief of the latter in reincarnation, he might be perhaps said to improve on his notion of the *ntoro* as a man's other and abiding self which remains near at hand so that he can hold communion with it. For, according to the Arunta theory, whereas one part or person, the *arumburinga*, is born afresh in each individual constituting what a schoolman would call his *principium individuationis*, its complement, the *koruna*, lives on from generation to generation, unchanging and immortal. Its place of habitation is the sacred storehouse where the bull-roarers are kept, many of them of enduring stone, such as are associated with names handed on through a succession of owners from ancient times. In this case too it is customary for one in need of spiritual support to seek out his deathless half, and drink of everlasting life as from some well springing within his own soul. But whereas the Ashanti *ntoro* is bound up with father-right and reflects a group-consciousness of a one-sided kind, the doctrine of the *koruna* treats the connexion with father and with mother alike as an accident. Each birth, or rather each quickening— for thus does the reincarnating spirit announce its coming as it pounces on the woman from the stock or stone where it abides with the other spirits of its particular totemic kind—is, as it were, an act of individual self-realization on the part of a unique being who, although his life is at once continuous and discrete, preserves enough sense of his identity as to be able in moments of religious concentration to bring his immortal and mortal natures—a Kantian might almost say his noumenal and his phenomenal selves—into a mystic union that yields or seems to yield the experience of a timeless present. One must apologize for reading into the mind of the Stone Age a philosophy at most subconsciously expressed in a few crude terms hardly detachable from the material objects that provide a cramping context for these beginnings of reflexion. Yet at its best metaphysics is but the poetry of religion translated into a rather stilted kind of prose ; and it is at least due to the Arunta version of the mystery of self-hood to point out that it accords in principle well enough with any modern theory of personality that seeks to do equal justice to its actual limits and to its large implications.

If now we pass on to Howitt's country, the south-east of Australia, with its many tribes dwelling on the whole in far more comfortable surroundings than do the Arunta of the central deserts, we appear to have leapt at one bound across the spiritual divide that separates a theological from a nontheological method of carrying on relations with the unseen side of the universe. Now Sir James Frazer, for whom this contrast is essential since he would found his distinction between religion and magic upon it, seems actually disposed to think that a richer environment is somehow responsible for this radical change of outlook. It is surely a new rendering of the parable of Dives and Lazarus that, whereas the former could afford to be pious, the latter must but console himself with superstition. As a matter of fact, if one surveys the institutions of the two regions in a wholesale way, it would be hard to show evidence of any marked difference in the general level of outlook ; and it might even be argued on diffusionist principles that the Arunta represent the spear-head of an intrusive movement from the north representing influences on the whole higher, in the sense of proving dominant wherever culture-contact is brought about. We must be careful, then, not to proceed off hand to class these primitive folk who happen to recognize some sort

of a god with all other types of theists as a united body
engaged in a straight fight with the entire body of the world's
atheists, including the unfortunate Arunta. On the contrary
we must attend exclusively to the special conditions governing the development of religion in Australia, so as to discover
whether the introduction of a god gives a new turn to ritual
and its interpretation, or whether on the contrary these
remain relatively unaffected in their broader and more
essential features.

Now it is to be noted in the first instance that the high gods
of the south-eastern region would appear to have the more
or less exclusive function of presiding over the initiation
of youth. It turns out, moreover, that the Arunta, despite
their godlessness, are in respect to the ritual of initiation
equipped with what is certainly the more elaborate type,
since it involves a fourfold graduation covering the greater
part of early manhood. On the other hand, there is nothing
in the available evidence—and it is as good of its kind as is
to be got from any part of the primitive world—to suggest
that the rite devoid of a presiding deity is from an ethical
point of view inferior, whether as judged by the precepts
actually imparted to the novices, or by the tone of earnestness
which so markedly characterizes the proceedings as a whole.
At most, perhaps, we can detect a slight difference in what
might be called the expert qualifications of the elders in
charge in the two cases. With the Arunta the ceremonies
are managed by just those ordinary or " lay " persons who
take the lead in tribal affairs, all of them, however, being
the more preoccupied with sacred matters the further they
advance in years and authority. Howitt's Yuin, however,
seem to have handed over the conduct of affairs to their
professional medicine-men ; for he certainly represents the
so-called *gommeras* to be adepts in such conjuring with
crystals as is confined among the Arunta to those who have
undergone a special training as healers. If, then, we regard
these mystagogues as a priesthood in the making, we are
perhaps on the track of an explanation why a High God
comes into existence, having the form of a high priest that

is lifted up into heaven and set upon a throne appropriately consisting in a huge crystal. In other words such anthropomorphism is specifically a magomorphism, an exaltation of the idea of the *gommera*. Nay, Daramulun, the god, shares with the *gommera* his title of biamban, or master of ceremonies ; and so intimate is the connexion between them that a human *gommera* can without offence play the part of Daramulun in person.

Whereas, then, the Arunta greybeard knows superhuman power and virtue only as manifested through those *churinga* to which he has access simply as a fully initiated member of the tribe, his counterpart among the Yuin is one of a select few whose capacity for ecstatic experience has provided the basis for a training designed to develop that capacity to the full. He too must have power or virtue, if he is to cure the sick, to make the rain fall, or to cause the youth to grow up into worthy men ; and, just as the Arunta who handles *churinga* in the proper spirit becomes himself " full of *churinga*," so the Yuin renders his experience of an inward grace as that of being " full of Daramulun." Something undoubtedly is gained in this change of symbolism, because it is more intelligible that a man should incorporate or impersonate a manlike being than a thing of stone or wood such as a bull-roarer. But primitive thought finds it hard to divorce the intelligible from the visible. Hence Daramulun must be, as it were, physically in evidence at the initiation mystery, so that he can be displayed to the novices as one who, as they are informed in his presence, can go everywhere and see everything, and will be pleased when they achieve the goodness that is expected of them. An image of a huge male with real crystals between his teeth, just as a *gommera* holds them there for all to see after pretending to bring them up from his inside, is either carved on the trunk of a growing tree or else built up in relief on the ground. Andrew Lang was perfectly justified in arguing in his *Making of Religion* that Daramulun with the rest of his congeners is no product of animism, but is, on the contrary, a " magnified non-natural man " as concretely anthropomorphic in conception as any Greek god. Moreover, he rightly insists that these Australian divinities, which have it as their main function to consecrate the moral education on which the continuity of tribal life depends, have an ethical character amounting to something spiritually high as contrasted with the low level of the material culture that accompanies it. Now all this is quite true as far as it goes. Yet, in view of the fact that the Arunta, while lacking the theism, nevertheless realizes the ethics by such means as some would stigmatize as fetishistic, one is bound to ask what religious advantage, if any, is secured by envisaging the transcendent power that helps to maintain the moral order in a distinctively human form. Now we need not treat as in itself an intellectual achievement of outstanding importance the mere capacity to construct the likeness of a man, and to attach a symbolic meaning thereto. The Palæolithic man of Europe could draw a human figure just as easily and well as that of an animal, if he chose. That on the whole, and with the curious exception of Eastern Spain where the taboo apparently did not hold, he avoided doing so must almost certainly have been due to a custom forbidding the practice because of its association with black magic. In other words anthropomorphism cuts both ways, since, instead of abetting reverence, it can equally well minister to the end of causing simple fear. Thus the Arunta turn to anthropomorphism only for the purpose of frightening the women with a tale about a hobgoblin who carries off the boys in order to eat them ; such a sop to female inquisitiveness scarcely veiling the threat to punish it with death, should it go the length of actually prying into mysteries intended for men only. Or, again, we hear from Queensland of a huge bogey-man, called *Nguru*, who, made of straw and furnished with a staring physiognomy as alarming as paint can make it, is solemnly burnt in the style of Guy Fawkes, apparently in order to scare the mosquitoes out of the country—an effect which a white observer solemnly avers to have followed *instanter*. Thus it is only fair to remember that anthropomorphism may in certain conditions actually run

counter to the spirit of reverence by suggesting various all
too feasible possibilities of misuse ; and it may be that the
prejudice against it in any of its forms, as entertained by the
ancient Israelite or by the modern Mahommedan—the relative
tolerance of the Christian being attributable to Gentile
and therefore ultimately pagan influences—is psychologically, and perhaps even historically, connected with the fear
that kept the cave sanctuaries of prehistoric France pure of
forms that might serve to deflect the incantations of the
ancient hunter from an animal to a human victim.

It might be argued, however, that so long as the worshipper
is at the mental stage at which he thinks as it were by means
of his eyes he cannot but derive religious benefit from the
fact that he finds himself face to face with a being sufficiently
like himself to be spoken to, and to seem to listen. If this
be indeed so, it is rather surprising that Howitt's south-
eastern tribes appear to be as universally innocent of
prayer as the Arunta of Spencer and Gillen. Some of us,
however, might consider that the Yuin are about to cross,
or have actually crossed, the dividing-line between prayer
and spell since, in the presence of the novices, the *gommeras*
solemnly dance to shouts of " Daramulun, Daramulun "
—the sacred name which must never be uttered on profane
occasions. Perhaps a neutral term such as " invocation "
is needed to cover such a use of a master-word, conceived
as it is to bring about a relation which would seem to be
one rather of sympathy or *rapport* than of converse by way of
the exchange of ideas. The use of the name is, in short,
incidental to the act of dancing Daramulun, so as to become
possessed by him, or, if we prefer to say so, by the power that
emanates from him. It is impersonation carried to the
length of an imagined transpersonation. Such a form of
religious experience is not easily brought within the psychological range of the civilized man, but it is reasonable to
suppose that the quality of it consists almost wholly in the
sheer intensity of the emotional tone ; and it may well be
that the Arunta derives from contact with his bull-roarer
a sensation that is altogether of the same order. One has

only to watch him taking part in a sacred ceremony, to see how it sets him all in a quiver. Now the Arunta is also fond of indulging in impersonation during the course of the initiation rites, but in this case he is simply dramatizing the legendary doings of his ancestors of the Alcheringa or Golden Age ; and these, though they occasionally display affinities with the totemic animals and plants, are on the whole human beings pure and simple, heroic rather than divine, though even so glorious, since, as the Book of Proverbs has it, "the glory of children are their fathers." Spencer and Gillen were unable to discern what precise value, of a wonder-working kind, if any, was supposed to attach to these pious observances, which, at any rate, were meant for the edification of the young, while the seniors no doubt enjoyed both as actors and expositors of tribal lore a pleasure at once æsthetic and intellectual. Even so these performances very possibly do not carry with them the rapture of the Daramulun dance. Daramulun, unlike the Alcheringa ancestors, has no history. He of whom the native can say in so many words that he goes everywhere and sees everything—though, to be sure, a competent medicine-man with his levitation, his second sight, and so on can boast to be almost as free of the limitations of space and time—has an infinite quality in which his devotees can possibly participate with more self-transcendence or, at any rate, self-abandon than one who merely loses himself in a whole-hearted rendering of saga. Without, then, pursuing further this comparison *in pari materia* between a theological and a non-theological mode of treating education in the light of a sacrament, let us say that, ethically and religiously speaking, the belief in a God makes for those concerned, if possibly some difference, yet almost certainly not much.

It remains to assess more positively the religious value of a symbolism expressing itself wholly or at least mainly in non-anthropomorphic figures of speech and thought. Now it must be assumed for historical purposes that even if some rare spirits may find satisfaction in an *amor intellectualis Dei*—though be it noted that in the eyes of his Christian contemporaries Spinoza ranked as little better than an atheist—religious value is for the majority of mankind by no means a thought-value in the first instance. Rather it would seem to consist in a suggestion of betterment gratefully accepted almost regardless of its source. It is the immediacy of the comfort vouchsafed that matters, and if it comes from all directions so much the better for humanity, always inclined to view its universe egocentrically. Given such a naive belief that the All exists ultimately for our benefit—and that is, perhaps, the fulcrum of the unconquerable will of Man—it is natural to conceive the environment, somewhat after the fashion of the Aristotelian cosmology, as bounded by a theosphere whence influence radiates through to the confused life of the earth from as many points as there are stars in the firmament. Recognizing the divine omnipresence, the savage tends to construe it in a distributive sense. He could subscribe to the text : " As for me, I will come into thy house in the multitude of thy mercy."

Thus it is now generally recognized that to render the *wakan* or *wakanda* of the Plains Indians of America by the " The Great Spirit," as the theological prepossessions of our missionaries formerly led them to do, is quite inadequate as a rendering of the diffused and, so to speak, adjectival character of a favour that distills out of surrounding space like dew upon the grass. Truly an old explorer might exclaim : " Their *wahconda* seems to be a protean god." The Christian missionary was startled, though perhaps inly flattered, to hear himself hailed as *wakanda ;* yet, when a native *shaman* arrogated likewise to himself that title, it obviously verged on blasphemy, more especially as he extended the same description to his songs, his appurtenances, and to his very tricks of sleight-of-hand. Another consequence which ensues from what a theology reduced to system and hence dominated by the logic of consistency must regard as a dangerous laxity of conception is that there is no authoritative doctrine to militate against particularism encouraging separate groups and even individuals to put their trust in a " medicine " of their private seeking. Such a method of trial and error, however, does not overlook the need of verification ; for it would seem that religious experiment is judged by its ethical fruits. Thus among the Omaha we learn from Miss Fletcher that tribal government is mainly controlled by a sacred order known as the Society of Night-dancers, admission to their ranks depending on being able to show a hundred deeds inspired by *wakanda*— " acts of grace," as she translates the native term which means literally " the causing of things having supernatural power." These consist in " acts and gifts which do not directly add to the comfort and wealth of the actor or donor, but which have relation to the welfare of the tribe by promoting internal order and peace." Provided with such credentials the members of the society find themselves *ipso facto* possessed of what other vocables containing a reference to *wakanda* represent as a direction of the divine energy, so that if it is " sent " against a law-breaker he is blasted—and the social ostracism thereby entailed amounts effectively to a death sentence—or, conversely, it can be " placed " on a good man who is in a tight place, so as to enable him to win through.

One might go on to illustrate from the rest of North America, where notions of the same type widely prevail, this approach to the divine by way of its multiplicity and diversity, such as is indeed suggested in our own *Benedicite omnia opera*. Take, for instance, the Algonkin *manitu* which is often used incorrectly to signify the individual totem— a sort of tutelary genius which is revealed in a dream to the fasting novice in the course of puberty, and befriends him ever afterwards in response to special attentions on his part ; this being in reality but a special application of a term of far more general import. So much did it trouble the explorers of the seventeenth century to impose a definite sense on so elusive a word that one of them, an Englishman, translates it " god," another, a Frenchman, prefers to render " the devil," while a third authority, also French, allows that it may stand equally for god and devil, since it was impartially applied by the Indians both to Europeans and to madmen. A modern account with deeper analysis notes that, whereas the grammar of the language makes a rigid distinction between the living and the unalive, *manitu* taken in itself and apart from the particular occasion of its manifestation is given the inanimate gender, as if it were a virtue or property attaching indifferently to things or to persons. It is further stated that this property is " omnipresent," so that " to experience a thrill is evidence enough of its existence." An example is given which it is hardly practicable for a civilized experient to try out for himself, namely, the thrill that is imparted by eating the heart of a brave enemy. A second illustration, however, invites a test more in accordance with our habits, if these include indulgence in a Turkish bath. For the prototype of this institution, vulgarly known as the sweat-lodge, exists here in order to promote godliness, rather than the cleanliness which is proximate thereto. According to the native theory " the *manitu* comes from its place of abode in the stone ; it becomes aroused by the heat of the fire ; it proceeds out of the stone when the water is sprinkled upon it ; and in the steam it enters into the body." Incidentally, it may be noted that by cutting oneself freely over the arms and legs the entrance of the *manitu* is much facilitated. As for proof, how else should it happen that " one feels so well after having been in the sweat-lodge " ?

Or, again, the Huron is prepared to credit anything and everything with its modicum of *orenda*, or mystic influence, the word meaning literally " song " and hence what might be described as a power of incantation. Thus, for instance, the cicada puts forth its *orenda* in order to ripen the corn by chirruping loudly in the morning when the day is going to be hot. The rabbit, on the other hand, sings metaphorically by barking the trees, thereby controlling the snow to pile itself up to that height and no further. Or a hunted animal can show that " its orenda is acute " by simply dodging the hunter ; who, on the other hand, is bound to catch it if his own *orenda* is greater. One is reminded of

what Professor Halliday has described under the name of
"the magical conflict," the principle of which is that it is
up to the wolf to see me first if he wants to get me. An
angry beast is said to be "making its orenda," the same
expression being used of a storm brewing.

But this crude attempt to envisage the world and its
denizens as instinct with a will-power hardly distinguishable
from brute-force and cunning cannot wholly preclude a
certain spiritualization of the notion of *orenda* when it comes
to stand for an inward experience rather than for the rule
proclaimed by external nature that every gambler in the
struggle for existence must stand by his luck. For we come
across two phrases at first sight presenting a contradiction,
yet offering between them what an analysis of the religious
attitude will disclose to be its true nature, namely that of a
bitter-sweet blend of fear and hope—of humility and
confidence. Thus, on the one hand, "he is arrayed in his
orenda" means that one is trying to obtain one's desire,
and hence is equivalent to saying "he hopes or expects."
On the other hand, however, "he lays down his own *orenda*"
stands for "he prays," indicating submission in the face
of a superior power. A man's *orenda* may be great, as is
said of a fine hunter or a wise *shaman*, and yet fate is stronger,
so that on occasion he must be ready to plead for his very
life. From religion he seeks a power of self-help, yet one
contingent on a crowning mercy ; and, whatever the
orenda that he can call his own, he knows that as measured
against an infinite *orenda* it is not enough, though there is
always more for the asking.

Passing, finally, to the Pacific region, whence comes the
notion of *mana* which has been adopted by the theorists
to give its name to a whole class of primitive concepts,
including the American examples just now discussed, one
could not dare in few sentences to try to bring out the full
force of Codrington's declaration that "all Melanesian
religion consists in getting *mana* for oneself or getting it used
for one's benefit." This much-quoted statement is indeed
by no means altogether a happy one, since it would seem

to treat religion as purely a concern of the individual—a contention that could not possibly hold true, or even half-true, of Melanesia or of any other part of the known world of Man. The point of the remark, then, consists rather in the comprehensiveness claimed for *mana* as capable of summing up the nature of the object to which religion—meaning, as Codrington goes on to explain, the whole body of religious practices, or, in a word, cult—has reference. But perhaps it would be better to substitute the word " purpose " or " motive " for object in this particular context, because what tends to defeat a civilized mind in its attempt to grasp such a characterization of the sacred or divine essence is precisely its lack of objectivity—an indeterminateness quite unsatisfying to those for whom Theology is a constitutional monarch having logic for her prime minister. It may be questioned, however, whether it be not but a form of idolatry to deem " the peace of God that passeth all under-standing " reducible to any system of thought-symbols ; since, just as the invisible does not lend itself to picturing, so the ineffable is not containable in any construction of words. Granted that what we have learnt to represent as a universe alike in its physical and in its moral aspect is for the Melanesian, or for any other savage, rather what William James would call a multiverse—a mere chaos of experiences, yet one not so void or formless that the spirit of God cannot be perceived as it moves darkling over the face of the deep. But there is a possible drawback in being able mentally to project a world-view as it were on to a screen, in that the onlooker is tempted to forget that if he were blind there would be nothing there for him to see. For that way lies the fallacy of imputing a pure transcendence to deity ; whereas of the two its immanence might be overstressed with less offence to the spirit of practical religion, which, historically at all events, would seem always to insist, if not on converse, at least on some kind of communion.

Let it suffice, then, in dealing with *mana* to note how, if it be, according to our notions, somewhat lax and fluid for purposes of objectification, that standing instrument of the

118

clear thinker, it nevertheless is not without a certain value
in the way of assisting religious consciousness, and one might
even say self-consciousness, on its subjective side. Here are,
for instance, some meanings given in Tregear's excellent
Maori-Polynesian dictionary to *mana* and its compounds.
In Maori it supplies us with words for affection, respect,
remembrance, exultation ; not to mention physiological,
or perhaps one should say psycho-physical, terms such as
heart, belly and breath, the internal apparatus, as it were,
through which *mana* works. Samoan, again, thus covers the
notions of desiring, loving, thinking, remembering, as well
as power to bless or curse. Hawaian, Tahitian, Tongan
recognize most of these senses, together with that of feeling,
or, again, of believing. Lastly in Mangarevan the word
manava stands for " the interior of a person " and is hence,
we are told, equivalent to " conscience " and " soul." If
the last word be a fair translation of the native idea, we have
climbed by means of a psychological ladder constructed of
more or less introspective material to the point usually
supposed to have been reached exclusively by way of
animism, with its tendency to picture the soul from without
as a *simulacrum* displaying what Tylor describes as a
" vaporous materiality." Of the two methods of reaching
the conception of an inner man there can be no doubt that
scientifically the *mana*-experience reaches the truth more
nearly than can any susceptibility to dream or hallucination,
however vivid. One must not exaggerate, of course, the
ethical quality of a sensitiveness to stimulation by way of
cult that is too wholesale and indiscriminate to permit of
any judicious selection among the influences that afford
the worshipper the partly real, partly illusory, sense of a
strong heart. Wherever he can tap *mana*, so to speak, he
drinks blindfold ; and, with little power of tasting to help
him to distinguish sound from poisonous liquor, rejoices
chiefly in the intoxicating effect, whether it serves to pull
him through a crisis, or is likely to incapacitate him for the
day's work. Thus, within certain limits, the intellectualizing
of religious experience is likely to purify it by providing an

immanent criticism ; which ceases to be helpful only at the point when thought tries to supplant will as the spiritual director of the whole man. Lest that happen, however, it is well to reflect on the history of religion in that youthful phase in which " the native hue of resolution " is by no means " sicklied o'er by the pale cast of thought " ; but, on the contrary, is the outward sign of an inward grace, consisting chiefly in an abiding faith that the world will always prove friendly to the man who struggles and seeks with all his heart.

6. RELIGIOUS ACTING

Starting as we do now on the third and final lap of our course, we are likely to find it the most trying. For we have to bring the feeling and thought manifested in primitive religion to the test of action. But is any religion, whether it has, or has not, reached the stage of accounting for its motives by means of concepts, going to stand up to that test, so long as it is administered by the man of science? Will he not insist on judging the efficacy of prayers for rain by their ascertainable effects on the rainfall? Will he not, in fact, rule out all symbolic action—and on the value of its symbolic practices the case for religion must stand or fall—as the pretence of action rather than the real thing? Let us see, then, whether the man of science need be quite such a philistine as he is usually painted.

To begin with, it is well to remember that anthropology, together with its chosen ministers, sociology and psychology, belongs to the biological, not the physical, department of the sciences. This is not the place in which to examine the difficulties special to his line of research which the physicist of to-day is so forward to acknowledge. Suffice it to say that he is more conscious now than ever before how his explanations conceived in terms of cause and effect leave him with an infinite series of conditions—something at which not only imagination but reason itself boggles. The biologist, on the other hand, makes an analogous use of function and end up to a point, but being thereupon faced by the prospect of another infinite series can at least put off the evil day by falling back on a category unknown to physics, namely, the category of organism. The actions of any living creature cannot simply be explained as an unrelated succession of movements, each in turn consisting of stimulus and reaction mediated by reflexion in proportion to the degree of conscious choice involved. For all these activities, over and above their separate workings, have a common meaning and value, in that they are alike contributory to a living system which is thus at once in them and above them. Thus the biologist cannot do without the cell as a minimum standard of reference whereby vital functions can be assigned their relative importance in some organic totality that pervades and owns them. Meanwhile, the student of Man can push biological investigation to its furthest because he has self-knowledge as well as external observation to guide him. He can thus supplement, or rather transcend, the category of organism by introducing the complementary notions of personality and society, each of which in its own way represents a higher unity such as affords a measure of the value of human actions taken in detail.

Now in his use of such notions the anthropologist tries to stand at one remove only from the facts of the time-process. Nevertheless it is essential for his scientific purpose that he should firmly plant one foot on this higher ground. Otherwise he can make no play with his favourite postulate of evolution, if it is to imply something more than process, namely, a progress according to some ideal scale. True, his criteria are not absolute. He makes a virtue of keeping in touch with sensible experience at all costs, and therefore leaves it to the philosopher and the theologian to see whether, without breaking altogether with crude fact, they can get away from it one remove further, and reach some standard unconditioned by time and change. From our present point of view, then, which is purely anthropological, it will suffice us to assume that there is a moral order just as necessary to success in our work, and therefore just as real for us, as the mechanical order in which human activities can be viewed in relation to one another. We need a system of double entry and cross-reference if we are to see a given action at once in its bearing on good living and on merely keeping alive. The test of survival-value, as it is usually termed, serves only to provide a downward limit of efficiency. In a swimming-race it is obviously no use to bestow appreciations of style and pace on those who sink and are out of it. But half the time when imputing value to survival we are surreptitiously adding to it the implication of dominance— of a capacity not only to live on but to live on more powerfully than before. This latter notion cannot have much significance in regard to the non-human forms of life, since a time-test applied to their extinct predecessors is the only available means of judging the relative staying-powers of those which are still in the running. In our own case, however, we have self-knowledge to help us ; and it must be part and parcel of our very faith in life itself that increase in vital power and an increased sense of vital power go together, not only in appearance, but in reality.

Nay, this inner conviction must be taken as more than a mere accompaniment furnishing a sign, like the rising of the mercury in a weather-glass. Our self-awareness likewise assures us that somehow we supply our own power, so that the sense of increase stands nearer to the ultimate source of it than the actual manifestation. In other words, the moral element in our actions explains them better than the mechanical element that links them with the time-process, because it reveals another dimension, that of an evolutionary or purposive order, which is as distinct from the order of mere sequence and coexistence as height is distinct from length and breadth. Hence we can boldly take stock of the material world and its opportunities in the light of what we will do therewith to forward our own self-realization. Moreover, history if interpreted from within shows that humanity, as represented alike by the individual and by society, has herein made measurable progress. So much, then, by way of prelude to a study of the value of religion in its active capacity—essentially a moral question, as the anthropologist no less than the philosopher or the theologian must view it.

By way of grasping the nettle firmly, then, let it be admitted from the outset that religious action, being symbolic, involves a kind of pretence, or rather would do so, if either the performer or the onlooker were to believe that what is suggested is carried out then and there in literal effect. From the standpoint, then, of the so-called practical man, which is but a specious name for the phenomenalist, the kind of thinker who treats the moral order as negligible, religion seems to be marking time instead of moving with it ; though why anyone should be the better for moving with it is more than his philosophy can unfold. At this point, no doubt, the psychologist will be inclined to interpose with his distinction between the extravert and the introvert, as if the normal man were not capable of being the two in one. It is true, however, that the rhythm of existence allows the hard business of keeping oneself alive to be punctuated with easier moments when the sense of being alive can be enjoyed for its own sake ; and that religion has always made the most of these intervals of spiritual leisure for its own purposes.

Though there is no reason why religion should not prove helpful in the thick of the vital struggle by hinting at a power in reserve, it undoubtedly cannot make its full appeal except to a mood that seeks solace from the worries of the daily round. In a word, its primary function is to be recreative. The etymology of our word holiday tells us as much. Or the fact comes out even more clearly if we study the calendar customs of the savage with their regular provision for what Professor Hutton Webster has well termed "rest-days." Almost as if it were at the dictation of the revolving seasons, which after all do directly determine the periodicity of the economic life, a corresponding cycle is organized so that the community in a body may withdraw now and then into another world, in which things in general and themselves in particular would seem to be more at their own disposal. For these annual "solemnities" have always stood with the primitive society for something deeper and more stirring than pleasant dissipation. In his half-symbolic, half-literal way—because, like the rest of us, he is apt to confuse the spirit with the letter—the savage feels that he is making the year go round, by a sort of encouragement which his rites impart to it no less than to himself. This is to turn the tables on the time-process with a vengeance; and I am by no means sure that it is not a truer version of the facts than to think of ourselves as chained to a cosmic machine as irresponsible as a runaway engine. To be sure that the little hills will hop in unison with one's own salta-tory efforts goes a long way towards building up the faith that will move mountains in a moral sense.

Let us, however, be ready to face the fact, such as it is, that the primitive mind has a tendency to project its feelings into things which, according to our physics, have none. It ought, therefore, to follow that the savage is bound to make a complete mess of his own physics. For instance, to treat one's spear as a friend and implore it to go straight shows a complete disregard for the mechanical principles of correct propulsion, and might therefore be supposed to end in a complete miss. As a matter of actual experience, however, it does nothing of the kind. The savage is no bad marksman; whether his invocation aids him or not, it certainly does not spoil his shooting. Apparently, while praying to God, he also keeps his powder dry. So, too, then, if one were to go steadily through the list of devices constituting his rather limited repertory in the way of applied physics, it is doubtful whether they would be found seriously the worse for being cked out with hortatory remarks directed more or less immediately towards his material or his tools. The psychological result, we may be pretty sure, would have been much the same if they had been addressed to himself. Self-identification with the object is the secret of an effective control of it, as all of us know who try to play games with skill. The more demonstrative Greek may exclaim " Be brave, my heart ! "—or at least the poet Homer makes him do so—but a Scotch caddie in like circumstances would be content to keep his eye on the ball. Meanwhile, as against possible cases, none of them very obvious, in which things have been positively mishandled by being treated as if they were sympathetic to human suggestion, there might be drawn up a formidable catalogue of useful discoveries made in the course of an experimentation conducted entirely on spiritual lines. It might almost be reckoned the leading principle of the early history of the arts that practical develop out of ceremonial uses. It is as if primitive imagination must be stirred by religion if it is to go ahead and invent ; whereas common sense, trudging behind, is only competent to select and exploit. After all, scientific curiosity is largely the lineal offspring of religious awe. Men lifted up their minds in fearful wonder to the heavens before they thought of making use of them in order to keep time or to steer a course.

Thus it is an utter mistake to regard the old-world weather-doctor as a humbug who does not clap on his ceremonies for rain until he is sure that the monsoon is about to break. On the contrary, having acquired as the humble servant and admirer of the monsoon a good understanding with it, he does his best to fall in with its ways ; such co-operation being out of the question had there been no love lost between them. Indeed, we may go so far as to say that a kind of introjection forms the psychological instrument whereby the mind creates objects out of the flux of experience. To be "itself," as we say, the object has to be invested with a certain selfhood or individuality. This process of objectification is intensified in proportion as the thing becomes a focus of emotional interest. Even the shipbuilder must not simply calculate in quantities of iron and wood, hemp and tar, but, if he is to be artist as well as mere mechanic, must enter into the very spirit of the design. As for the sailor, all his pride in his vessel, and therefore three-quarters of his seamanship, will be due to the fact that he regards her as a mistress. This anthropocentric habit, then, is bound up with the very possibility of an intelligent grasp of the universe, which otherwise could be no more than a multiverse, if even that. Idealized, that is to say moralized, to the full, the anthropocentric attitude becomes the theocentric. Thus the savage is but obeying the fundamental impulse on which all advance in science as well as in the rest of our culture depends when he reads a sort of double of himself into things as a prime condition of becoming further acquainted with them. By staging a diplomatic approach to them, as his ceremonies enable him to do, he gives himself a chance of framing a preliminary intuition of them, of reading their dispositions in their faces, that cannot but stand him in good stead when he comes into closer dealings with them. The sooner one gets the hang of the characters, the better one is likely to follow the working out of the play.

When therefore Nietzsche in his wild way declared, " Action has no sense ; it merely binds us to existence," he was but stating what would be perfectly true on a denial of the validity of the moral order. Action wasted on things that are neither good nor bad but simply indifferent, as abstract matter is by us conceived to be, would be an aimless shifting of dirt. If action is to count, it must somehow minister to our scheme of self-cultivation. Its value must depend on the degree in which it helps to further man's huge design of domesticating his universe—of bringing the human family together into one communion, so that their very surroundings may reflect the warmth of their hearts. Perhaps the primitive group, consisting of a score or two of souls in loose alliance with other groups totalling at most a few hundreds, would be more ready than one of the vast economic aggregates composing nations of to-day to realize that their moral relations constitute the secret of their luck ; nor are their material successes so great as to blind them to the futility of trying to live simply by them or for them. For this very reason, then, they naively extend the hand of friendship to natural agencies quite incapable of returning the compliment in a literal way. Thus they are undoubtedly guilty of what is known as the pathetic fallacy just in so far as they allow themselves to be taken in by it. In other words, we with our superior knowledge of physics have a right to accuse them of meeting the situation with the wrong kind of action just in so far as they mean to produce a mechanical effect and actually think that a non-mechanical means will suffice for the purpose. On the other hand, so far as this is not the intention, the charge falls to the ground. Nay, the fallacy is transferred, as it were, to the objector, if he fails to see that it is perfectly just and reasonable for the poet or the mystic, instead of holding the mirror to Nature, to hold Nature up as a mirror to himself so that he may become aware of his emotions in their imaginary reactions which are nevertheless real reflections of his mood. To repeat, then : to impute the pathetic fallacy where there is no self-deception is a fallacy of mal-observation to which the naturalistically-minded are especially prone. Because the extraction of sermons from stones does not accord with their laboratory methods, they ignore the important part played by such a process in the larger workshop of the moral life.

The anthropologist, then, must beware of mistaking for error what is rather a kind of nescience on the part of the savage. Thus when Hobhouse accuses him of " confusion of categories," it may be well asked how he could possibly confuse categories of which he has never heard. For to insist that, if he had conceived them, he might have used them to his own advantage is neither here nor there. As it is, he somehow manages to do without them. He tends not only to talk but to think in holophrase, attaching its meaning to the sentence without taking separate note of the verbal elements that compose it ; and therefore the latter are not words at all in our sense, but something else—radicals, functions, or whatever we like to call them. So too, then, he does not feel the need, as we should do, to disjoin the parts of a coherent vision of the various outstanding objects of his little world which has at once a moral and a mechanical aspect, the former predominating in proportion as the object as a whole touches his interest nearly. For he makes it his prime business to size up in a general way the value of the things about him considered simply as associates. On the strength of such a valuation he has to decide on the degree of intimacy that is possible and desirable between himself and them. What they are means what they can do for him, and in this *mana* of theirs is summed up all their properties in a sort of *qualitas qualitatum*. Hence when we in our critical way want to know if the *mana* is personal or impersonal, whether animism is implied, and so on, the savage is perfectly entitled to reply, as he will if pressed, " let it be as Your Honour pleases " ; for such distinctions are quite foreign to his method of character-reading as applied impartially to men and things so that each may be assigned what one might call its " luck-index."

Not that there is no regard for matter of fact, as we should distinguish it. Savage vocabularies usually turn out to be especially rich in names for the different animals and plants. Indeed, a heart-to-heart talk with a game-tracker about the beasts of the forest or with some old wife about the simples that she culls for medicinal purposes will show that much accurate observation is mixed up with a good deal of romance ; the latter being too loosely conjoined with the common-sense element to bring about felt contradiction, and to lead to conflicting and therefore ineffective reactions to the same practical situation. Some of what is here termed romance really deserves such a description, amounting to no more than the wonder-tale which is passed from mouth to mouth to enliven a dull hour. This concession to the play of fancy, however, is not taken seriously, and at most may help to foster the spirit of credulity by pandering to

I

the taste for marvels. On the other hand, the beliefs that are largely responsible for the attribution of *mana* are likewise of the romantic order in the sense that a certain sympathetic imaginativeness enters into their essence ; but at the same time they stand for truth in its highest degree, as it were a truth by revelation. For the *mana* of anything is its esoteric meaning imparted only to the initiated and as a tremendous secret. Correspondingly, then, it represents the inwardness of the thing, the hidden reality behind its appearances as they are for the vulgar. Indeed, we may be pretty sure that in the first instance it is no ordinary and " once-born " person, but some sage elder credited with supernormal powers by his fellows, who by means of his second sight, so to speak, brought into view the *mana* or " occult quality " in a given case by giving it expression in some appropriate ceremony. Thereupon it would take a certain effort on the part of the community to retain its sense of the special significance of the object so distinguished and, as it were, consecrated ; and it would be the duty of a succession of mystagogues to hand on the sacred tradition, no doubt with additions and variations of their own. In this way the average man on receiving such instruction would be vaguely conscious of the difference between the sensible properties of the thing and the ritual value or over-meaning constituting a sort of *aura* visible only to the eye of faith. In like proportion he will be aware that his actions in regard to it belong to two planes, the lower of which is concerned with mechanical effects, while the higher is reserved for manifestations not so much caused as granted, sheer bounties on the part of a Nature teeming with goodwill towards those who are in touch with her.

Take the case of sacred stones. On the face of it nothing could be more unresponsive to blandishment than a stone, or on the other hand more submissive to mechanical treatment such as hammering and chipping. From the dawn of history Man has displayed a wonderful command over such material, and might consequently be supposed to labour under no illusions as to its physical characters. Nor does he in regard to what he reckons to be common stones. But an uncommon type of stone intrigues him. If, for instance, it is shaped like a yam, he conjectures that this cannot be for nothing, and therefore plants it in his yam-patch to see whether it has *mana* for yams, taking care to explain to it exactly what is wanted. If the crop is not signally improved, the stone was but a common one after all. If on the other hand it works miracles for him, it does not follow that it will go on doing so when he dies, but his son who inherits it must try out its inherent power afresh. Or a friend may wish to acquire it from him at a price; but what he really pays for is the accompanying formula of invocation whereby the original owner established communion with the stone, the goodwill of which becomes transmissible only in so far as the ceremonial routine is maintained. Or, again, a stone is like an animal, though it is not forbidden to use a little chipping to improve the resemblance. Such a " figure-stone "—and, whether Boucher de Perthes was right or wrong in making it a feature of Lower Palæolithic times, it is certainly known to the Pueblo Indians of to-day—forms an excellent hunting-charm if the ritual adjuncts are not neglected. For instance, one must suck the *mana* out of the stone and emit it in the form of the hunting-cry of that particular animal. Or, again, one must not forget when the quarry is slain to dip the stone in the blood to rejoice its heart; for this is alive though the rest of it is dead. One might give many other examples of stones that owe their *mana* to their queerness—the meteorite, for instance, which may have called men's attention first to the virtue of iron; or the crystal, which must come from the Rainbow because his colours are in it, while it can also throw men into a kind of trance if they are made to stare at it.

Of a rather different, though no less characteristic, order of sanctity are the conspicuous stones that form landmarks, as they also do for those who set up no such god of boundaries to delimit their private holdings. Thus for the Central Australians they constitute a standing road-map whereby the legendary wanderings of their totemic ancestors can be at once memorized and authenticated. No wonder that the reincarnating spirits of these same ancestors use them as lairs from which they can invade the women that they severally choose to be their next mothers. Indeed, the native is never off haunted ground, and, as his initiation proceeds, becomes more and more preoccupied with what is for him at least the sheer historicity of his surroundings. Living a miserable life if we assess it in terms of its creature comforts, he nevertheless succeeds in glutting himself with his sense of kindliness towards those bare rocks which from time immemorial have served his people for a home, and can bear silent but perpetual witness to their glorious doings. When one hears of savages objecting to the blasting and quarrying operations of the white man who is opening up their country, they may be going a little far in supposing that their resentment to the desecration of these natural shrines of traditional sentiment is shared by the very stones themselves, who will surely bring some judgment to pass. In any case, however, the sin of sacrilege is not definable in terms of mechanics or of a positivism founded on mechanics, and no one can say that the primitive version of it, which feels with the holy object rather than about it, seriously misrepresents the moral issue. No doubt there may be awkward corollaries from the mechanical point of view—as for instance if it be held unlawful to build a sacred edifice with other than unhewn blocks. Even so, taken in its religious context such a taboo may be quite defensible ; the test being whether it is an article necessary to the faith whereby those concerned are encouraged to make the best of their lives.

Or, again, take the case of the primitive man's relations with the animals and plants of his environment. Here we by no means find the poetry at cross-purposes with the prose so that he gathers or hunts any the worse for being a totemist. One might expatiate at great length on the ingenuity of his methods as a tracker and trapper. One is amazed, for instance, at the Red Indian method of catching eagles for the sake of their feathers, so much prized for ceremonial reasons ; when the hunter, lying flat on his back in a pit covered over with branches, on which the lure is placed, actually grasps the stooping bird by the talons—a procedure involving even more skill and courage than the rather simpler practice followed from time immemorial by the snarers of hawks on passage at Valkenswaard in Holland. Or, as regards plants, what could be more masterly than the way in which cassava bread is prepared from the poisonous roots of the manioc—a risky proceeding for which the South American Indian deserves all the credit, unless some of it is passed on to the natives of Africa for so quickly and successfully adopting the art when introduced by the Portuguese.

How comes it, then, that a technical skill implying so nice an appreciation of matter-of-fact should be conjoined with a great deal of etiquette quite inconsistently implying that the behaviour of animals and plants can be swayed by paying them suitable attentions ? The answer can only be that, as we may verify in our own case, mankind can bear with inconsistency to an amazing extent by packing away their sentiments and thoughts in separate logic-tight compartments. Just so one finds the keen sportsman and the sympathetic naturalist conjoined in the same person, who finds little or no psychological difficulty in making his game-bird the object of gun or camera according to season. For at the worst the mind can persuade itself that what it chooses to differentiate in theory must likewise be disjoined in fact. Thus the totemite can contemplate with equanimity the slaying of any number of specimens of his namesake animal, so long at any rate as this is done by members of other totems, for whose express benefit he himself does his best to secure an unlimited increase on the part of the species ; but that the species as such can suffer detriment in the sufferings of the individuals that compose it is a view that conveniently escapes all notice. On the contrary the species is held to enjoy a lot superior to the mischances of this mortal life, being of course conceived according to no definition of a species that a modern biologist could adopt—

if indeed there exists any that will satisfy him—but rather as the folk-tales represent Brer Rabbit or Reynard the Fox, namely, as a sort of deathless Super-rabbit or Super-fox reminding one of the " concrete universal " of the Hegelians. To such a well-spring of life the pitcher may go back as often as necessary without prejudicing the bountiful supply ; and the human heart may well feel gratitude towards the power manifested in this general willingness to provide individually unwilling victims of economic necessity, as Man sees it from his angle. So too the Cornmother or John Barleycorn is an immortal, whose most serious disability is a certain tendency to shape-shifting. As Andrew Lang makes the spectre of Castle Perilous explain : " What we suffer from most is a kind of *aphasia*. . . . We don't know how we are going to manifest ourselves." Even for the practised and calculating agriculturist of the pre-scientific era, the actual seed may die and nevertheless quicken again by virtue of some divine incorruptibility immanent therein ; while to the mere gatherer, for whom the Australian wilderness puts forth an unlimited store of grass-seed in due season, the *mana* or sheer miracle of it stands out distinct and sublime as apart from the tiresome and unedifying business of purveying a daily sufficiency for hungry bellies.

There remains to be considered what from our point of view may seem to be the most self-contradictory of the attributions of divinity to objects of the sense-world—namely, the case in which a mere human being is regarded and treated as more or less a god. It will be found, however, that on the whole religious practice, which with the savage counts for so much more than doctrine, manages to keep the human and divine aspects sufficiently apart to prevent heavenly from being degraded to the level of earthly standards. Divinity in whatever degree it be recognized remains an ideal character but loosely attached to the human occasion of its manifestation, who is thus rather vehicle than owner. A typical example is afforded by the supernatural power assigned in general to the dead. Every living man, according to the primitive view of the matter, is capable of one day becoming a ghost, and as such affecting the friends and enemies that he leaves behind him for better or worse. On the other hand, there are few traces, if any, of the modern conception of a withdrawal of the disembodied soul behind some impenetrable veil, to await that final reassembling which is likewise to be a day of reckoning for each and all. In the society of a static type, when custom changes so little from generation to generation that a theory of reincarnation is almost demanded by the logic of the situation, the living and the dead have interests so identical that there is every reason why they should remain in touch. During a certain interval of time, indeed, each individual ghost, since men die singly, demands particular attentions from those of his own kin ; and, while they for their part are cut off from the rest of society by their uncleanness as mourners, so the dead man in his turn hovers between the place of spirits and his former house, as divided in mind and ambiguous in status, and consequently as perplexed and prostrate, as any of his friends. Let once the period of mourning be overpast, however—as is often marked by the ceremony of a second or " dry " funeral—and, as it were by a completed initiation, the liberated soul is enrolled in the general host of the departed, and has achieved what amounts to a fresh age-grade, giving him a certain precedence in authority even over the generation of elders that is supreme among living men.

There is no doubt a certain inconsistency in combining any form of manes-worship, logical outcome though it be of any tendance of the dead, with a theory of palingenesis, since no trailing clouds of glory can altogether obscure the fact of the helplessness of the infant condition. But primitive metaphysics at this point falls back on the philosophic device known as a hypostasy—in other words, an abstraction promoted to substantive rank. Souls, say the Arunta, are really double, consisting of an immortal element associated with another that is subject to continual re-birth. The former part fortunately haunts that holy of holies on earth where the sacred bull-roarers are secretly stored ; and here the living man can commune with his higher self by renewing the bond between his twin constituent principles, so that something undying seems to be transfused with his mortal nature. We are told by Spencer and Gillen that, as the native advances in years, he becomes more and more preoccupied with sacred matters, and spends most of his time in a state of what amounts to inarticulate meditation ; though the place of actual thought-forms acting by means of words as vehicles of religious emotion is taken by motor processes, manipulations, choric movements and so forth, that stir the heart directly with effects comparable to those of music. The civilized man, however, for his part has no ear for this music and hence can make little or nothing of its wordless message. How, then, can we estimate the spiritual value of ceremonial practices, even when shadowy beliefs can be cited in support of them, which have no clear analogy within the religion or even the philosophy of the Western world ? That a man should somehow feel stronger and richer for imagining himself a two-in-one, a being that in its human capacity can reach out so far as to embrace a trans-human soul, hypostasis, person, aspect, or whatever this better half of his veritable self is to be termed, is not to be dismissed offhand as an aberration, a discarded superstition, because our type of man is for the moment interested in other matters. Certain it is that, if the science of comparative religion is to march in step with anthropological research in general, it must once for all reject the postulate of a unilinear evolution. The great adventure is something more than an omnibus ride along a highroad. Thus on the opposite hypothesis of a differential progress of mankind, an advance along a broad front by isolated rushes though on the whole conforming to a direction, it is possible to credit certain developments of historical religion with a meaning of their own, quite apart from their bearing on the course of any of the several so-called world-religions that compete for the attention of the more or less civilized peoples.

In the present instance, then, we must be careful not to

deny the religious import of this primitive version of the
" transcendental ego " because it is independent of theology
in the literal meaning of the word. A belief in God and a
belief in personal immortality naturally seem to us to go
together, though as a matter of history it might perhaps
be shown that this association of ideas is rather late,
and coincides with that stage of thought when a retribution
theory replaces what Tylor calls a " continuation theory "
of post-mortem existence. Be this as it may, a personal
interest in the fate of one's own soul stands by itself as the
only major motive of religion at all equipollent with the
recognition of a *Deus sive Natura* as at once dispenser of
material benefits and maintainer of the moral order. It is,
however, usual to suppose that at the level of savagery the
individual has scarcely found himself : so that any personal
religion must be out of the question. Now it may be true
that no salvationism of the type fostered by the mystery
religions of classical antiquity and likewise represented, and
perhaps echoed, within Christianity, has any close counterpart among primitive folk, whose particular line of religious
evolution leads them rather towards an intensification of the
tribal system. Even so, as the Stone-Age culture of Australia
abundantly testifies, the very fact of having a name, with a
bull-roarer in lasting stone to vouch for it, is enough to
encourage in the individual the beginnings of self-consciousness, and, to judge by his communings with his immortal
double, the beginnings of an effort to enlarge and perfect
his inner man. One might go on to note how the so-called
medicine-man, whose greater susceptibility to ecstatic
experience has marked him out for his special vocation, uses
his personal totem to enforce his claim to wonder-working
power, and at the same time rejoices in the conscious possession of so intimate a guide and protector. Indeed, the whole
subject of that phase of religion known to anthropologists
as " nagualism "—the *nagual* of Central America being
equivalent to the personal *manitu* of North America and
presenting a certain resemblance to the tutelary genius of
the Ancient Roman—deserves more attention than it has

hitherto received. Whereas, thanks largely to Durkheim and his sociological school, the moral significance of primitive religion has been illustrated and explained simply in its relation to society as an ideal category, the co-ordinate and in some sense rival category of personality has been largely overlooked as furnishing a no less supreme object of striving throughout Man's religious history. It may even be that, just as self-sacrifice and self-realization are ends which ethics fails to reconcile completely so long as our human limitations are kept in view, so religion as historically conditioned must always hesitate between more or less incompatible notions of deity, the one transcendent, or at least trans-personal, like the State, the other immanent and as it were the *alter ego* of the pure and pious soul.

Turning now to the social aspect of religious experience, the function of religion is in general that of investing the human leader of men with an authority not his own, but superhuman in that it relieves him of ultimate responsibility for his mistakes. At the savage level there are no philosophers ready to criticise the world-process or the moral order ; but on the other hand there are grumblers in plenty among the lesser folk as soon as things go wrong, and, as it silences doubt, so it inspires unlimited confidence that an infallible power should be regarded as the real head of affairs. Not that we must think of tribal society in terms of the alleged class-war of modern times, namely as an exploitation of the masses by the more cunning and therefore wealthy minority, in whose hands religion becomes but a disguised instrument of tyranny, a specious means of inculcating a *Sklavenmoral*. On the contrary, there is only one type of despotism known to the primitive community, which takes the benevolent form of a régime of strict custom. In such conditions it is perhaps easier for the many to acquiesce in the control of the few, when the latter are plainly no less subject than the rest to ancestral precedent—something as impersonal yet all-pervading as the very air that they breathe. After all, so long as culture is propagated by purely oral means, the repositories of tradition, the men as it were of the longest memories, must always have the last word when it is a question of bringing public behaviour under the right bye-law. Moreover, such an unwritten book of the rules makes due allowance for the fact that life is uncertain. Even crisis repeats itself with sufficient regularity for a routine to be devised wherewith to meet it ; and, although the individual undergoes the shock of novel experience at every stage of his vital career, and even the group may be taken unawares by visitations ranging from war to pestilence, or from eclipse to earthqake, there will always be someone old and wise in council—in most cases, indeed, there will be a whole senate composed of such persons—who can prescribe some time-honoured means of mitigating anxiety and encouraging a return to that cheerful outlook which is, fortunately, normal to Man's happy-go-lucky disposition. Meanwhile, the man himself, however prominent he may be during certain hours of his little day, is recognized to be but a mouthpiece of the sacred lore of the tribe. He is " full of *churinga* "—bull-roarers—as the Arunta say ; or " he has a great *ertnatulunga* "—is a veritable storehouse of such mystic objects. Here we have the idea of inspiration translated into stone-age metaphor. Note, however, that when such a man grows senile, he is said to have " lost his *ertnatulunga*," and his friends are shabby enough to give him a less brilliant funeral, because his very ghost must share in this manifest decline of power. So, too, at a later stage of political development the divine king must die as soon as his faculties show signs of waning. And yet in his time he may have been hardly distinguishable from the great Nature-powers, the Sun, the Thunder, or the Rain, with which by virtue of his office he held sympathetic communion causing Earth and Sky to do their duty by his people. Perhaps, unlike the Australian grey-beard whose glory has departed with his usefulness, so that he goes down dishonoured to his grave, the potentate who is slain in order to preserve the royal *mana* intact may have the consolation of contemplating himself as in some sense a dying god, entitled to a shrine of his own in the temple of tribal remembrance. Even so, the man in him must die in order that the divine part may be liberated.

Once more, then, we find it to be the function of religion to emphasise the moral at the expense of the material aspect. Just as the sacred stone or the sacred animal is not sacred *qua* stone or animal, so the sacred man is sacred, not as mortal man, but as a symbol of something undying—in the case of the king, the undying life of society. Religious action is efficacious just in so far as it serves to disengage the ideal meaning from the gross imagery in which the earthbound human spirit is bound to express itself. Personality and sociality between them, or possibly some absolute reconciliation of two principles that up to a certain point are obviously complementary, stand for moral truth as eternal as thought can make it ; and the supreme object of religious faith cannot be the vulgar moving of any material mountains, but rather the moving of the human mind, of the inner man, so that it may be turned right round from the materialism that goes with sheer animalism, and thus may envisage and pursue moral perfection in the shape of individuality and community jointly purified, enlarged, and, in a word, made real. Such, I venture to suggest, is the lesson to be drawn from the history of religion as a record of the evolution or unfolding of the profoundest purpose manifested in human life ; and, though my business is not to play apologist for the savage, but simply to account for the facts about him, I ask anyone who is at pains to study the evidence whether the will to organize self and society until they meet and merge in a divine goodness is not provably as deeply-lying and widespread as human nature itself.

PART III.

PRE-THEOLOGICAL RELIGION: PARTICULAR ILLUSTRATIONS

7. RITUALISM AS A DISEASE
OF RELIGION

$\mathbf{B}$y ritualism I shall understand the overstressing of the ritual element in religion; meaning thereby its use of prescribed forms, its routine. The word is often used in English in this unfavourable sense by those who object to what they consider to be an exaggerated formalism in Christian worship. Such an attitude looks back to the Pauline distinction between outward sign and inward grace, between letter and spirit. That, of course, is a matter of theology, which does not concern us here. But it affords a verbal justification for enquiring, under a convenient title, how far human religion in general exhibits a similar tendency towards a formalism noxious to its successful functioning as a normal activity of Man.

We cannot, however, attack this particular question without raising a larger one, namely, whether a pathology of institutions comes within the range of anthropology at all. How can a science which aims at being objective, that is, non-confessional, non-tendencious, say that some social developments are more healthy—in other words, better—than others ? Would this not be a value-judgment rather than a judgment of fact, a purely existential proposition such as alone befits a science of the empirical or positive order ?

The usual reply to this would be that anthropology as a branch of biology has the right to use the test of survival ; because a differential mortality—one that eliminates one kind of life in favour of another—appears to be true in fact, and is therefore objectively traceable in its effects. Granting all this, one may, nevertheless, go on to ask whether in practice such a test will work for the purposes of Social Anthropology.

I seem, indeed, to remember in Montesquieu a statement to the following effect : " The Troglodytes systematically violated their contracts and so perished utterly." Give me enough of such facts, duly authenticated, and I will undertake to explain the causes of the rise and fall of nations. But no ethnographer or historian can produce them in reference to the peoples that are dead and gone, much less if the folk that they study continue to exist. One must recognize a certain immunity from the stress of natural selection in that biological dominance which enables the whole human species to live with a considerable margin of safety in its competition with other species ; while intra-specifically the same thing happens, so that the more powerful the group the heavier handicap of vice it can afford to carry. Thus, on application to human affairs, the survival-test breaks down.

As a *pis-aller*, however, an expedient may be tried which at all events has the authority of Aristotle behind it, and involves the use of the comparative method. We can consult τὰ λεγόμενα—what people actually say about the working of their own institutions. If on their own confession they are dissatisfied with them—if they allow that the shoe pinches—then is not the evidence strong enough to warrant the anthropologist in assuming that something is wrong ? This, then, might be called the test of normal function. Human needs happen to be so much alike everywhere and always that one general pattern, reflected in the social organization, will be found to fit the lot. On the other hand, the means taken to satisfy these needs vary considerably, and thanks to experience, which is experiment, improve ; so that we can broadly distinguish grades, which are likewise stages, of culture, implying an increasingly satisfactory mechanism for supplying needs that in their profounder psychological characters have changed very little, if at all. Working on these lines, then, let us see whether in the particular case before us this method will prove helpful, even if it be not above criticism as a way of reaching objective results. In defence of it, however, let us remind ourselves that we are not laying down values in the style of the moralist or legislator, but are dealing with actual valuations which, when once made, become fact—in other words, matter for history. We study human nature, which includes a no less natural faculty and exercise of choice.

It remains, then, to attack the specific question whether ritual—the use of set forms in religion—can be overdone, our method being to enquire how far the religious consciousness thereby finds itself disappointed in its expectation. To a like extent we may presume aberration from normal function, which is to serve the will to live.

We must begin by discovering the permanent need to which religion corresponds ? That there is such a need may be inferred from the universality of religion as a human institution. Clearly the subject calls for more discussion than is possible at the moment, and we must be content with a rough generalization. Let us say, then, avoiding all question of the truth of the beliefs on which it depends —for that is outside our subject—that religion is an art of self-encouragement in the face of the uncertainties of life. It is as it were an extended system of faith-healing covering activities of all kinds in so far as they exceed the bounds of calculation. Such a purely psychological account of it must suffice us now.

So much for the need of a confidence which must be some-how created despite all appearances to the contrary. As for the organization of the confidence-making art, it would seem that, historically, it arises as a by-product of all manner of activities of a critical nature, being incidental thereto like the general's exhortation before the battle. Gradually, however, this occasional character gives way to one that is periodic, and correspondingly the art has a chance of developing an independent technique which comes into the hands of experts.

Given, then, such an art in process of evolution in the only way open to human nature, namely, through experimenta-tion, it is not likely to start with any just notion of its capacity and limitations, any more than the faith-healer can tell except by trying what ills of the flesh he is competent to cure and what not. Thus, whereas experience all along testifies to the general helpfulness of a bold attitude in the confrontation of danger, such as overcomes doubt, even though it be in itself reasonable, and paves the way to a strenuous counteraction, it is not easy at first for Man, with his very small acquaintance with mechanical aids, to distinguish between the moral and the physical powers severally brought into play when his energies are thus invigorated. Just as he is apt to read his skill, so he reads his faith, into his instrument when the former rather than the latter is the determining factor. Thus he mistakes for a supermechanics what is really a superethics. The " uplift," as they say in America, is projected into the material part of the work done, though it is wholly germane to the spiritual part.

With fuller self-consciousness, however, enlightenment on this point is bound to come. The spear with *mana* which rendered its original owner invincible is transferred to a weakling, and behold ! the mystic power is cut off at the source. So too, then, with formulæ. They may seem to be endowed with automatic efficacy so long as the right man uses them, but for the wrong man they manifestly refuse to work. Whereupon it becomes incumbent on the religious consciousness to tackle the question : Which is the right man and which is the wrong, who thus severally can and cannot command the inherent virtue of the sacred words ?

This process of discovering a moral justification for the effective use and enjoyment of religion would doubtless be quickened if religion were wholly an individual affair, since as between one man and another the relation between merit and good fortune tends to be fairly obvious. But religion, more especially in its more primitive phases, is mainly communal in its appeal ; and for a whole community to be convinced of its own unworthiness there must be some very strong pressure, such as that of imminent disaster— a precarious condition, however, to which the small society is more constantly liable than the big. Meanwhile, so long as prosperity reigns, the social self-satisfaction will tend to be such as not to invite too nice an enquiry into the efficacy of a religion conducted with a purely external propriety. A more or less thought-proof traditionalism will satisfy those sound conservatives who would reform nothing, and least of all their morals.

In such circumstances there can hardly fail to be slackness in the actual conduct of the customary rites. They will come to seem tedious and expensive. The purely liturgical expert with his tiresome pedantry cannot supply the want of a moral discipline which can affect the hearts of simple men. Thus the Scribes and Pharisees are always the butt of the moral reformer, who on a conviction of sin founds a new faith in religion as a force that works inwardly.

Automatism, then, as imputed to ritual stands for one of two things, either a careless acquiescence in things as they are, or a survival from that primitive state of mind which projects the spirit of an enterprise into its physical accompaniments. For the latter fault there can be no excuse when self-consciousness is well enough developed for the springs of moral action to have become reasonably plain. If we use the word " magic " by way of antithesis to religion as a similar use of symbolic action differing in respect to the anti-social motive involved, then, although I would hold it religious in the savage, who knows no better, to be inclined to externalize the helpful quality in his rites, I yet would brand as anti-religious and therefore quasi-magical the same tendency as exhibited in one of the higher religions. For these at most can tolerate an outward conformity in their humbler followers, since the leading minds cannot fail to be aware that rites are not causal but educational agencies—that what is wanted of them is to change, not circumstances, but hearts.

In this connection it may be noted that, whereas there is nothing to be said for the view of magic that would roundly identify it with the use of concrete symbols, as if these could imply nothing but a materialistic conception of the power set in motion by religion, nevertheless undoubtedly language as the prime minister of thought helps greatly to promote a fuller consciousness of its supreme function as a stimulus of the soul. When the grosser forms of symbolism survive into the higher religions, as typically in the case of sacrifice —in itself a crude affair of shedding blood and appropriate only to those who do their own butchering—it needs much verbal reinterpretation to allegorize the primitive act. A service of prayer, on the other hand, in spite of all the stiffening in the way of vain repetitions which ecclesiastical tradition may import into it, embodies a far more direct appeal to the awakened intelligence, and at the same time is far more easily detected whenever it degenerates into nonsense. No doubt religious thought must always remain imaginative, dealing as it does with ultimate mysteries ; so that the logical element in the language of religion must always be supplemented by a sensuous element. Indeed, it typically displays a quasi-poetical character and is associated with music, or, earlier, with dancing. Thus a religion is always more than a moral philosophy in its hold on the practical life, because its emotional accompaniments serve to bring the meaning home to the mind as a whole. Among these accompaniments tradition positively counts as a consecrating influence, since men, for reasons both good and bad, have always been immensely susceptible to the glamour of the past. But respect for tradition must not interfere with the continuous cultivation of that immanent energy which is the life-blood of religion. Apart from this nothing remains but the empty shell, a mere curiosity fit only for the shelves of an anthropological museum.

8. THE SACRAMENT OF FOOD

In the fifth chapter of Miss Jane Harrison's *Themis*, which sets out to deal with the triple subject of totemism, sacrament, and sacrifice, the central interest of the treatment is revealed in the appended quotation, " What meanest thou by this word Sacrament ? " Starting from the *omophagia*, or eating of raw flesh, that survived in Greek ritual of the Dionysiac order, Miss Harrison professes her physical repugnance to such a crude proceeding, but promises that our normal repugnance will disappear when the gist of the rite is understood. Thereupon she tries to prove that the *omophagia* was, in its primary intention, neither a sacrifice pleasing to a personal god, nor a sacrament in which communion with such a god was brought about. Originally, she argues, the rite in question was " part of a system of sanctities that knew no gods," belonging as it did to a " social organization, namely totemism, that preceded theology." She goes on to show, here echoing Durkheim, that totemism essentially embodies a collective form of experience, being based on the supposed relation between a human and some other natural group, such as typically an animal species. This relation, as she is able to prove from the Australian evidence, is explicitly conceived by the totemite as a kind of identity. " That one is just the same as me," says the kangaroo-man of his eponymous animal. She might also have cited his even more telling phrase that the kangaroo is " all-one-flesh " with him. Now flesh, together with the accompanying and even more mysterious blood, may well stand for kinship, more especially when considered in what was presumably its earliest form, the tie between mother and child founded on a real, and not merely symbolic, community of flesh and blood. Miss Harrison, however, is more immediately interested in the blood-relationship imputed to the namesake animal or plant. Now this is clearly a symbolic or non-literal connection, even if we make all allowance for the tendency of the savage mind to " confuse its categories," as the late Professor Hobhouse puts it—or, let us say less technically, to blur its distinctions. Paying attention chiefly to one type of totemic rite, the so-called *Intichiuma* ceremony for the multiplication of the totem, she interprets it as an attempt on the part of the human group to realize their oneness with the animal group ; the dawning consciousness of a difference between them being transcended in such an effort to bridge the gulf and lose themselves in a higher unity, a wider communion. Hence she terms such a rite, for all that it takes the form of a dramatic representation of the desired identity, " methectic " rather than " mimetic." In other words, instead of imitating something external and to that extent alien, they seek direct participation in a consubstantial nature that is thus not theirs and theirs at once and together. So far, then, as participation prevails over imitation—in other words, so far as undifferentiated thinking and group-emotion predominate—we may have sacrament and even sacrifice, but the net result is only a kind of magic. For Miss Harrison defines magic as " the manipulation of *mana*," which, according to her conception of it, is itself an undifferentiated kind of power or grace proceeding from sacred things in general. Religion, on the other hand, in her view implies worship, which can only be towards a power invested with a separate and distinctive being of its own—in a word, a " god." Let this brief account of Miss Harrison's general position suffice ; though the by-paths through which she wanders while finding her way to these conclusions are such as to defy summary notice.

Now what I am about to say is not meant as an attack on Miss Harrison's views, if only because in many respects I agree with her line of thought, and simply desire to suggest sundry modifications of the main argument. These are the more necessary because so much fresh evidence has accumulated since *Themis* was written—that is to say, in the course of the last twenty years. On one major issue only do I definitely disagree ; and, since it is largely a question of nomenclature, I had better clear up the matter at once, lest an unacknowledged difference of terminology lead to misunderstanding later on. I need hardly say that the stumbling-block consists in the familiar puzzle how to draw the line between magic and religion. Now Miss Harrison appears to me to be trying to run both with the hare and with the hounds. Sir James Frazer may be taken to represent the hare, while I can claim for myself a humble place in the pursuing pack. Thus on the one hand she accepts the Frazerian view that there is no religion until there is propitiation of a personality with superhuman attributes—a conception of religion that appears to me to be based on the principle, " Orthodoxy is my doxy." On the other hand, by connecting sacrament with magic, she deprives religion of one of its chosen instruments ; unless it is possible—as I do not think it is—to hold that magical bottles are suitable for preserving religious wine in all its integrity. Probably she is here following the lead, not of Sir James Frazer, but rather of Robertson Smith, the foremost pioneer in regard to all these questions ; for he was ready to class magic as just that early stage of religion when ritual seems all-sufficient because its motives remain subconscious. Indeed, Sir James Frazer was content to accept this opinion when he started

to write *The Golden Bough*. Only in the second edition did
he take the fatal step of dissociating magic altogether from
religion on the strength of an erroneous psychology that
identified the former with a kind of sham science founded
on the abuse of analogy—as if the same could not be said by
its detractors of the most advanced theology. Miss Harrison,
however, shows no sign of postulating an age of magic
separated from the age of religion by a hiatus that is logical
or chronological as you please. On the contrary, her rather
inconsistent language would imply that magic is religion at
its pre-theological stage ; and she actually describes a system
of sanctities that knows no God as " a form of religion." If,
then, it is to be Frazer or Robertson Smith, let it by all means
be the latter. I protest, however, in the name of historical
continuity, against the use of any such invidious distinctions.
For surely it is invidious to use a term of disparagement such
as " magic " to imply that the cult of the sacred is inferior
in religious value in strict proportion as dogma is absent.
Besides, when one looks closely into a given case, there
usually turns out to be more doctrine of a kind at the back
of the ritual performance than might be supposed by those
who forget that thinking is a good deal older than logical
thinking. I doubt indeed whether any theology would be
wise to depend in the last resort on its logic. Even so it has
to be freely admitted that the doctrinal background of the
Intichiuma rite needs an interpreter of dreams to fathom its
meaning—a meaning none the less in which a mystic might
possibly find more satisfaction than in a creed reduced to
Aristotelian terms. To close this part of the discussion, it is
surely better to treat sacrament as a fundamental institution
of religion from its first appearance in history onward to its
fullest and most refined manifestations. For by so doing we
shall be less likely to do an injustice either to the savage by
consigning him to an isolation ward where he must play at
his idle mysteries by himself, or to the votary of the most
advanced religion by accusing him of following a way of life
that is encumbered with dead matter.

Being prepared, then, to extend the name of religion to the

kind of rite which, though godless, nevertheless belongs to the cult of the sacred, or, as Miss Harrison would say, to a " system of sanctities," let us proceed to find a meaning for the term " sacrament " such as will be not inconsistent with such a pre-theological attitude on the part of the believer. Now it should be stated at once that neither Miss Harrison nor I would for one moment wish the food-rite—the Intichiuma ceremony and its like—to be regarded as the only kind of sacrament. She makes this quite clear by saying, " There are other means of contact, of sacramental communion, besides eating and drinking." But the commensal meal, as Robertson Smith would call it, is at least thoroughly typical of the ritual proceedings to which totemism may give rise. As for its sacramental character, this consists in Miss Harrison's opinion in being a means of what she calls " *mana*-communion." Again, though less happily in my view, she speaks of sacrifice as " magical contact," evidently using the word " contact " as equivalent to communion. In other words, a sacrament is a special means of establishing connection with what she calls a sanctity. Such in the case of the Intichiuma ceremony is the sacred totem ; the object of the rite being to enable the worshipper to participate in that sanctity—to share in a *mana*, or miraculous quality of abounding kindliness and goodwill, which has been in a way common to totem and totemite all along, but it is now stirred into greater activity by inducing a fuller consciousness of its real presence. Miss Harrison's main point in laying stress on the participatory aspect of the rite is that she wishes to preclude the idea of a favour conferred from without, as by a god who of his own initiative rewards his worshippers either according to their deserts or beyond them. In the case before us, on the contrary, the *mana* is immanent in the rite itself, if we include the parties to the rite among the conditions of its realization. Thus it is like some meeting of lovers whose awareness of each other is quickened by coming into touch, so that their warmth becomes a flame. From an objective point of view, no doubt, there are two parties to such a transaction, but subjectively, and in terms of their feelings, all difference is for the time-being transcended. Such a relation, then, must be classed as sympathetic rather than contractual. In short, it is a matter of joining souls rather than of marrying a fortune. In putting it thus I am perhaps straying rather far from the letter of my text ; but I do not think that I am doing an injustice to the general purport of Miss Harrison's thought.

Such, then, by hypothesis being primitive sacrament in its root idea—namely, a rite embodying a blessing that is enjoyed rather than imputed, spontaneous rather than derived—let us go on to see how far this interpretation is borne out by the known facts about the Intichiuma ceremony. It is, however, only fair to Miss Harrison to say that she does not regard the Arunta custom as altogether characteristic of that age of primal innocence of which she is in search. I am afraid that with us anthropologists the original condition of this or that institution always tends to lie just over the horizon ; for Nature, unlike logic, abhors absolute beginnings. Now the primal group of communicants —the first *thiasos*, as Miss Harrison would say—would for her undoubtedly be the matrilineal kin already possessed of a totem wherein its community of blood is made known to itself. Now, when such a group eats together, it may well be supposed that a certain festal warmth would invade and enhance their mutual relations. Yet surely the last thing such a group would eat with comfortable abandon would be their totem ; unless indeed Miss Harrison is ready to accept Professor Haddon's theory that originally, that is to say, just before observation becomes possible, the totem was the staple food of the group, so that the kangaroo-eaters became known to their neighbours and eventually to themselves as the kangaroo-men. If, however, we stick to known facts, a totem is not so much the first as the last thing on which a totemite would be inclined to make a hearty meal. The totemic relation, in fact, is normally expressed rather by fasting than by feasting. Hence one would have somehow to argue, if the satisfying effects of a full meal are insisted on as the emotional basis of the rite, that the kin-group has by further evolution become so sophisticated that it has learnt to get an additional thrill out of breaking a taboo —that a soupçon of sin can impart zest to participation in a sanctity. For the moment, however, it will be enough to note that the food-rite as we find it in actual practice is pervaded with a certain shyness or shrinking—in a word, with the taboo feeling. In short, the postulated association between the sacramental and the festal is not borne out by the example cited. So far is the Intichiuma ceremony from being an *instantia crucis*, that the arm of the *crux* or " sign-post " points in quite another direction.

A second objection to the use made of the Arunta food-rite as a clue that is to lead us back to the commensal meal of a group owning one and the same strain of mothers' blood is that these totemites are classified by the anthropologist as " cult-societies " pure and simple. It is true that membership depends on a sort of birth-qualification ; but so peculiar are Arunta views on the subject of birth that one can only term the imputed relationship metaphysical, since it rests on no physical basis whatever. A man's mother may be of one totem and his father of another, and yet he will belong to a third, if his mother was entered by a reincarnating spirit when passing the stock or stone inhabited by unborn spirits of that third denomination. Now this may be a survival of a very primitive type of totemism, as Sir James Frazer believes it to be ; or, as others hold, it is the mark of a peculiar, not to say eccentric, elaboration of the older system that associates totem with natal group in the ordinary sense. For our present purpose, however, we need merely note that a group of Arunta totemites is not a kin in the social sense—in other words, is not a natal association that could have ever been brought up together so as to rejoice in a common life and more especially in common meals. A kangaroo-man's sleeping and eating arrangements have nothing to do with his totemic status. Thus the chosen example of sacramental communion falls so far short in a second respect of answering to the supposed archetype that one almost begins to wonder whether the Arunta stand any nearer to the imaginary beginning of things than do Miss Harrison's Kouretes.

Let us see, then, whether we can get at least a little way behind the Intichiuma rite by treating it not quite so strictly as a development of totemism. After all, if there has been any decided change of opinion in anthropological circles during the last twenty years in regard to religious origins, it has been all in the direction of reducing the part played by totemism as such during that early phase which is conterminous with the hunting and gathering stage of society. On a closer scrutiny it has been found increasingly impossible to keep the specifically totemistic features apart from the non-totemistic in those observances, more or less common to the whole wild-feeding world, that are meant to bring about luck in procuring their daily food. I do not think that " zoolatry " is a particularly happy term to express this attitude of the hunter towards the live things hunted ; though it is better than " theriolatry " which ignores the plants altogether. But, if we do not lay undue stress on the implication of worship lurking in the termination " -latry ", it will perhaps do. Indeed, my own prejudice is in favour of stretching the word " religious " to its utmost so as to cover practices which are of dubious validity as well as manifestations of the religious spirit at its purest and best. Thus even when conciliation rather than control appears to be the leading motive, one is never quite sure that such a term as cajolery will not cover both intentions alike—a cajolery, however, which co-exists with, and in a way is based on, a very real respect. A wise savage might well aspire to control a vegetable or one of the milder animals, and yet would surely prefer to conciliate a dangerous beast, though none the less in the hope of getting him in the end. Mark, too, that in a very real sense the most unpleasant to tackle of all the denizens of the wild may be regarded as good ; for, as Miss Harrison proves from a number of primitive vocabularies, " good " and " eatable " may start as more or less convertible terms. Thus, although his character for beneficence is as it were thrust upon the dangerous beast rather than of his own seeking, his flatterers almost mean what they say when they laud his excellence to the skies ; not being too clear in their own minds whether they are referring to his succulent taste or to his kindly disposition. Besides, there is another reason why an animal or a plant —for this consideration applies to them all—should be treated nicely lest something go wrong with the hunting. For no sooner is one member of the species killed and eaten, than another is wanted in its place, so as to be killed and eaten to-morrow. There is needed some supernatural suspension of the law that you cannot eat your goose and have him. How precisely it is to take place is hardly man's concern, one would think, so long as it does take place. Nevertheless, in expressing his desire in the collective gesture which custom prescribes he usually manages to convey a hint to Providence how the thing might be done. The simplest method is to send off each individual animal, as it is put to death by the hunter, so that it may come alive again as soon as it conveniently can. Thus by leaving some part intact, the eater gives the remaining parts a chance to reassemble ; so that, for instance, he will consume the flesh, but must abstain from cracking the bones. It is obvious, however, that there is much to be said for a more wholesale way of inducing a given species to keep up its numbers. Nay, whereas replacement one by one can merely maintain the *status quo*, there is no saying how glorious a hunting would not ensue if the animal kind as such could be made to devote its full energies to breeding. This of course is exactly what the Intichiuma rite tries to do ; and the fact that the performers have to be namesakes to the given animal or plant is perhaps a secondary matter. For in reality the Arunta community as a whole is ultimately responsible for its system of rites of multiplication. But, having at its disposal a set of animal and plant identities—not to say names—distributed in a rather arbitrary way among its members, it naturally chooses as its go-betweens those persons who are in special touch with the various species contributory to the tribal food-supply. Thereupon the specialists are allowed to deal in their own way with their totems' private mode of multi-plication—sometimes a very complicated affair, as in the case of the Witchetty-grubs, of which both Spencer and Gillen became the adopted relatives. It is certain, however, that tribal celebrations of great importance—such as the *Engwura* or Fire-Ceremony, which lasted some four months—are pervaded from end to end with totemic performances, even though these are always enacted by the separate groups, who, while as it were retaining the copyright, must yet say their piece for the edification of all. Whatever, then, be the far-off origin of the Intichiuma rite—a highly speculative matter at the best—its actual function, taken at its widest, is, I contend, to further easy finding and plentiful eating on behalf of the tribe in general.

Now, if this be so, the eating of the totem—by no means a universal feature of these rites, and prominent only in a few —must be subordinate in function and meaning to the motive operating throughout. As Miss Harrison is quick to see, sacrifice and the food-sacrament go closely together. How to slay and eat nicely, so that the animals will not mind, is the problem that they have to solve between them. How, then, do the occasional parts of the Intichiuma rite that involve killing and eating help towards this solution? As regards the alleged commensality whereby the totem is entertained by his fellow-totemites at his own expense, in the first place I cannot find the slightest evidence that any such idea is present in the minds of those who perform the rite, nor can I see precisely how it would fit in with what the occasion really requires, which is surely an apology. On the other hand, I believe that if we never lose sight of the fundamental fact that the envoys are ambassadors between the tribe and the animal or plant whose name they bear, the rest will explain itself. These ambassadors, then, are there to express on behalf of the whole Arunta people a hope and a fear.

Firstly, then, as to the hope. This hope is for food in plenty. The leader of the Witchetty-grubs rubs the stomachs of his whole party with a stone symbolizing the grub and exclaims, "You have eaten much food." This is, as Miss Harrison would say, a *dromenon*, a thing done in the sense of done ahead. It prefigures the desired abundance. If it were meant as a feast, it would be a poor substitute for the real thing, since grub inside the stomach and a stone applied to its outer surface are satisfactions that will not readily fuse. But to give utterance in the most vivid gesture-language to the common hope by means of the prefigurement of its miraculous realization—such might well be the mission of envoys who had no private and sinister interest in forwarding the petition.

Next as to the fear. When the kangaroo-men actually kill a kangaroo and each eats a morsel, and only a morsel, thereof, are not the totemites in this case taking on themselves by anticipation the brunt of that invidious action on which the tribe proposes presently to engage on the widest scale? It is the principle of handselling—of leaving the costly first step to those most competent to take it with impunity. There is direct evidence that the totemite by eating a little is held to make such eating free for the rest. Thus so far from being of the communial type, the rite is incipiently piacular. The function of the mediator is to take upon himself the sin of the rest, so that the righteous indignation of the victimized totem may be deflected to the one group of human beings whom he would be most disposed to forgive.

Now I do not wish to exaggerate the contrast between the communial and the piacular type of ceremony, though the symbolism of sharing a blessing will naturally differ a good deal from that of getting rid of a curse. For the fact that they have likewise something in common is indicated by the very word that in English has come chiefly to stand for a rite of reconciliation presupposing the removal of a cause of trouble —in moral terms, a forgiveness of sin. When we say "atonement" our very pronunciation helps to conceal the etymological sense of achieving an at-oneness. At-oneness, however, is by no means the same thing as oneness. It stands for a transcended duality rather than for a unity that is such by nature. So, too, then, I suggest that the so-called communial rite is from its first inception intended to effect a miracle of at-one-ment—to build a supernatural bridge across a natural divide. Now Miss Harrison, I suppose, as well as Robertson Smith—whom I take her to be following closely whenever she is not taking her lead from the very similar speculations of Durkheim—would say that the original kin-group has its internal squabbles to get over, so that the common mother's blood of them would need warming up from time to time if peace and friendliness were to reign in the primæval home. I grant that stern taboos were presumably needed to preserve the amenities in the earliest of fire-circles. Thus the prohibition against intestine murder and intestine marriage may alike have originated in the attempt to enforce orderly and seemly relations under the sanction of the curse of the united mothers, the high priestesses of the common blood. There is no evidence, however, so far as I am aware, that their custom of eating together —after all, a daily occurence which for this very reason would be unlikely to acquire any special significance and sanctity—had developed into a means of repressing dissociative tendencies, whether in the form of a quarrel with a kinbrother or a liaison with a kin-sister. The only fact I know that seems at all to the point—and it is very sketchily reported by Howitt and may well be incorrect—is that certain totemites ate an erring brother apparently by way of restoring him to their communion. As, however, endocannibalism can never have been the normal mode of supplying the common table of the archetypal kin, this solitary instance, even if its documentary value were above suspicion, would hardly be relevant. Killing and eating a blood-relation once in a blue moon—for a group that needed to do this at regular intervals would not remain in being long—could hardly serve as the starting point of a custom embodying the idea of commensality. In any case it clearly involves reconciliation rather than spontaneous good-fellowship as its basic motive. Hence I would not seek in the internal affairs of the kin for the reason that makes eating together a symbol of communion, but rather in its external relations with the stranger. That, if hospitality be given and accepted, the foreigner's potency for harm will be neutralized, is a principle holding throughout the savage world. But from the first such a precept implies at-one-ment—a duality to be transcended. I fully allow that the symbolism implies that eating together in fellowship, as the kin normally does, is incompatible with the spirit of enmity ; so that the stranger, by being treated as a kinsman, is miraculously transformed from a foe into a friend. But it is only at the point at which symbolism becomes necessary that ritual comes into existence. The commensal meal of the kin may have provided the pattern for the rite ; but in itself it is no rite—any more than, if, for religious reasons, the horns of a bull are worn by a man, the wearing of horns is proved to have been a religious act on the part of the bull.

If, then, I am at all right in insisting on the fundamental disharmony which the food-rite, whether in its distinctively totemic form or otherwise, is designed to overcome, the food-animal, totem though it be, is not like a kinsman participating in the common meal by natural right, but is rather like the alien who, if persuaded to accept a gift of food, cannot in common decency maintain the malignant attitude affected by foreign devils as such. It is surely a matter for diplomacy—for the most delicate handling of an international situation in which the one party is trying to get the best of the other with as good a conscience as it can muster—to persuade a poor beast that it has to be killed and eaten, because that is how the world is made. When the thing is settled—the one-sided treaty signed—there may be rejoicings on the part of the eater ; but to force this aspect of the matter on the attention of the about-to-be-eaten would be bad policy and bad manners to boot. In short, the occasion calls rather for the ceremonial lamentations that so often accompany the death of the animal, even when it has become domesticated, that is, has sunk to the level of a member of the *familia*, the group of farm-chattels. Now one does not like to think of such a show of grief as simply a piece of solemn humbug. For mankind is so honestly convinced of his innate superiority that he cannot conceive the under-dog to have rights, if these run counter to interests which, pertaining to so high a being as himself, must surely be valid absolutely and for all alike. We can call it the imperialistic fallacy, if indeed it is a fallacy for strength to vindicate its natural dominance over weakness. Be this as it may, primitive man is either too cautious or too decent-minded to adopt the tone of a bully in his dealings with those shy creatures, the game ; but is always polite, even when he does not mean quite all he says. Indeed, when one thinks of the very primitive hunter as essentially a man of snares— for he has not got weapons for successful attack in the open —one is impelled to ask whether his very ritual is not a sort of supersnare. Facts, alas ! are facts, even if one would prefer not to find an organized hypocrisy among the roots of religion.

Let me refrain, however, from further speculation in this unedifying direction, lest I prejudice my case. My main contention is simply that in origin the sacramental type of meal is essentially distinct from the festal, even if later religious practice sometimes tends to confuse them. The festal type is certainly to be encountered within the wide ambit of historical religion ; which includes, for instance, downright Saturnalia, when the ordinary, or profane, use of food helps to bring people together and promote general hilarity. But, in connection with the sacramental type, no such festival spirit is endurable. As a well-known verse from the Epistle of St. Jude proclaims, " These are spots in your feasts of charity, when they feast with you, feeding themselves without fear." In the true sacrament a holy fear must preclude free enjoyment of the food as such. There need be no eating at all, as when the flesh and blood are consumed by fire ; or, if there is eating, it must at least be sparing. For it is an act not so much of consecration as of deconsecration. Not participation but naturalization is its object. Man is not so much concerned to add to his own stock of *mana* by absorbing that of the victim, as he is to neutralize an alien *mana* which might do him harm if, like the blood of righteous Abel, the victim's blood should continue to cry aloud for vengeance. The analogy, in short, is more with a composition negotiated in a matter of blood-revenge than with any collegiate gathering—any sort of primitive bump-supper. To repeat what was previously said, the sacramental meal is not so much a feast as a fast. Nay, it often demands a previous discipline of abstinence, its ascetic quality being perhaps emphasised by the use of an emetic. It is, I would suggest, the presence of this taboo-feeling—this shrinking from the world and its pleasures —that helps to dematerialize the sacramental meal, and thus to render it the fitting symbol of a communion with the divine when approached in a truly religious spirit, namely one that must be humble before it can become jubilant. Miss Harrison, in her clever way, perceives that in the actual cases analysed there is always a certain felt difference between the man and his totem-animal. But she thinks this sense of apartness is secondary—a faltering of the original all-embracing sense of kinship ; whereas, if I am right, it is primary, nay, integral to the very conception of the sacramental act.

The topic being inexhaustible I could go on to argue that another point made further on in the same chapter by Miss Harrison, to the effect that the idea of sacramental communion is in the middle religions supplanted by a theory of " *do ut des*," assumes a new appearance if we identify communion with a sort of peacemaking between potential enemies. For it will be found that the savage notion of giving and receiving presents is by no means the same as our own, being largely symbolic and, in fact, embodying another characteristic method of patching up a quarrel If so, the transition from communion to the gift-sacrifice is not so abrupt as Miss Harrison would suppose.

But, instead of pursuing this new hare, let me break off at this point, with apologies for having sounded a note of criticism without at the same time expressing my genuine admiration for Miss Harrison's work ; for I owe much personally to the inspiration derived from her writings. Yet after all no greater compliment can be paid to her as a serious anthropologist than by subjecting her theories to critical examination. Reading her *Themis* over again carefully, I seem to perceive some slips in the details, but feel at the same time that in the greater matters she exhibits an unfailing sense of direction together with an unfaltering resolution to win through. In fact, as Andrew Lang once said about her, she has *l'étincelle*.

9. RELIGION AND THE MEANS OF LIFE

To be uncertain about the whence and the when of the next meal is an unenviable state of mind, more especially if circumstances tend to make it a permanent condition. Yet this is precisely the preoccupation that haunts the daily and nightly thoughts of the savage who remains at what is known as the food-gathering stage. He must take up his spear and go after game that shows no desire to be caught, while his wife, basket or wooden trough in hand, must scratch the soil for roots or scour the bushes for berries or even grubs, both in season and out of season. It is by no means unknown that whole families should starve to death.

For it is not as if the human race had clung to the warm and plentiful zone that must have furnished its original habitat. On the contrary, it is found making the best of things in all the most inhospitable environments, even though still dependent on the scraps that chance throws in its way. Indeed, the animal world presents no other example of so widely ranging a creature ; and nothing but the confidence proceeding from a high brain charged with immense if chiefly latent powers could have justified this reckless dispersion, extending through long periods of time and from the Equator to the utmost verge of the Polar seas. Now faith is but another word for confidence, and it needs faith to conceive a deity and pray to him for help. Such an attitude would always be classed as religious. But when the hunting savage puts his trust in rites that, however sacred in his eyes, do not clearly imply relations with any kind of personal power at the back of Nature, we are apt to set them down as magical. Yet so rigid a distinction cannot but obscure the fact that, alike in motive and method, the two procedures have a great deal in common.

Man's object being in either case to move the assumed friendly influence to work wonders on his behalf, he tries to do so by using some means, whether word or gesture, of suggesting exactly what he wants. How precisely the meaning is conveyed or how it makes itself felt is a question on which the primitive mind is quite unable to frame a clear opinion. But that it somehow helps in practice to relieve the situation in times of crisis is known and appreciated as a matter of direct experience.

Nor is the efficacy of spell or prayer tested simply or even mainly by the extent to which the behaviour of the external world is actually observed to be swayed. It is rather the inward effect—the conviction acquired of succour at hand— that heartens the hungry folk so that they are fed on the very hope of food, and a little miraculously stands for plenty. What, then, is popularly described as magic is really religion in the making, if instead of stressing the difference between savagery and civilization one insists rather on the continuity of the process of development. Passing on, therefore, from what, after all, is largely a question of words, let us study in the concrete some examples of rites of the economic type, taken both from the food-gatherers, who, though of prehistoric type, can be found in existence in various parts of the world to-day, and from their more advanced brethren the food-producers.

Anyone who has had the luck to visit one of the painted caves of France or Spain will have noticed the javelins affixed to the flanks of the game animals there depicted, together with numerous other signs showing that this was no mere art for art's sake, but served some symbolic purpose. Not the faintest whisper has reached us, however, of what the prehistoric European believed when he held his mysteries in these dark and silent places. On the other hand, the Stone-Age man who survives in Australia still carries out ceremonies that in spirit, if not in detail, must be similar enough to afford an instructive analogy.

In the barren centre of the Continent the leading interest from a mystic point of view consists in encouraging the animals and plants to multiply, and incidentally in encouraging mankind to look forward to this consummation. Thus the witchetty-grub is ceremonially stimulated by the witchetty-grub men, who bear that name and deem themselves one in kin with the insect in question. Their headman, in order to induce it to lay eggs, makes solemn bows over a shield decorated on one side with wavy lines to show how former grubs walked about in ancient times, and on the other side with circles, the larger ones representing the bushes on which they fed, the smaller ones being their eggs. This is done once a year just a little before their breeding time. When the insect has obligingly increased in numbers, then the grub man, who must otherwise religiously abstain from eating his namesake, has to partake of a small portion in order to retain his power of persuading it to be fertile ; after which the rest of the tribe, on whose behalf he has exercised his magic powers, are free to take their fill of the delicacy.

Again, the emu men for like reasons make a sacred design, but this time it is on the ground, a plot being cleared and levelled, then drenched with their own blood to form a hard surface, and finally decorated with black, white, yellow and red paint. The drawing is not realistic in the style of

the cave art of Europe, but, on the contrary, is highly
conventionalised ; so that, for instance, two large yellow
patches stand for lumps of emu fat, of which the natives are
very fond, other smaller patches in yellow are the eggs in
the ovary, and so on.

Once more, the kangaroo men trust to the symbolism of a
painting, though preferring to use a rock face for its ground.
This, too, is very far from imitating nature, consisting as
it does in little more than rows of vertical lines, some white to
represent bones and the rest red to indicate the fur.   The
performers of the rite solemnly open their veins and allow
the blood to spurt over the ceremonial stone.   Being " all-
one-flesh," as the native phrase puts it, with the sacred
animal, they renew their communion with it through this
sacrificial act.   Those who choose to speak of magic in such a
connexion should at least be ready to recognize here the
first beginnings of religious practices and beliefs of the most
profound significance.

Passing from Australia to North-eastern Asia we find nature-
fed tribes, evidently long established there and in many
respects reminiscent of the ancient hunters of the far past,
who indulge in like observances with intent to conciliate the
game, having devised this method of keeping up the live
stock before ever the art of domestication had brought the
problem down from the spiritual to the material plane.
Thus bear-meat appeals to the Ainu, who must therefore
see to it that the bear does his part to maintain the supply—
a thing that he is not likely to do if he is not kindly disposed
towards the Ainu.   So a baby bear is secured and for a time
nurtured with the greatest solicitude, the women actually
giving the infant animal the breast as if it were one of their
own children.

At last, however, comes the day—the cub having become fat
and succulent in the meantime—when it must die, or,
rather, must go back to be reborn ; and this is obviously a
good opportunity of asking it to take a message to the bears
and generally to the gods assuring them of the love—a
cynic might call it the cupboard-love—of the Ainu.   " The

Ainu who will kill you," they explain, "is the best shot among us. There he is, he weeps and asks your forgiveness ; you will feel almost nothing, it will be done so quickly. We cannot feed you always, as you will understand. We have done enough for you ; it is now your turn to sacrifice yourself for us. You will ask God to send us, for the winter, plenty of others and sables, and, for the summer, seals and fish in abundance. Do not forget our messages ; we love you much, and our children will never forget you."

Or, again, among the Otawas of Canada, the Bear clan—who, after all, were bears, and so might well enter into the animal's private feelings—made him a feast of his own flesh, addressing him thus : " Cherish us no grudge because we have killed you. You have sense ; you see that our children are hungry. They love you and wish to take you into their bodies. Is it not glorious to be eaten by the children of a chief?" Surely no reasonable bear ought to mind when treated so handsomely, more especially if, as often among the American Indians, the bear was no sooner slain that a pipe was lit and put into his mouth, while after he had been eaten his head was hung on a post and complimentary orations were lavished upon it. And so it may have happened ages ago in prehistoric Europe ; for in the cave of Tuc d'Audoubert the head of a mighty bear had been set high upon a pillar of shining stalagmite, while numerous foot-marks in the clay floor testify to the presence of congregations long departed.

Again, America can provide other hunting rites that possibly throw light on the very remote history of our race. It was no less an authority than Boucher de Perthes, who was the first to prove palæoliths to be the genuine handiwork of man, that brought forward certain figure-stones, as they are termed, from the same layers as had yielded the flint implements which he had just vindicated ; and these, too, he claimed to be of human workmanship. Indeed controversy rages to this day over the question whether these animal-like forms roughly shadowed forth in battered stone are merely freaks of Nature or the symbols of some vanished

faith. Be this as it may, the making of figure-stones not unlike
in appearance is something that appeals to the primitive
mind to-day.

Thus among the Pueblo Indians of Zuñi in New Mexico
six kinds of hunting animal, the eagle, the wolf and so on,
are held to preside over the six regions of the world and
all the medicine powers contained therein.   Hence it
is extremely lucky to find a stone shaped like one of these,
or one which by artificial means can be made to resemble
it. For the beast of prey through its image can control the
game which is it wont to hunt, thus enabling the hunter who
carries the charm to do the same. Now, though the rest of
the animal has become stone, the heart is alive within.
So the hunter puts it to his mouth, draws a deep breath, and
presently emits the hunting cry of the powerful beast, which
so daunts the game that a kill is ensured. The hunter,
however, must be careful to dip the stone in the blood,
bidding it " drink, that it may enlarge its heart." The act
undoubtedly enlarges his own heart.

It might be thought that the domestication of animals
would once for all rob them of their sacred and quasi-divine
character. For, if a wild boar excites respect and even awe
in one who sets about its capture, the same feelings can
scarcely be entertained towards the pig in its sty ; though
the uncleanness so often imputed to it is really a mark of
holiness. But old notions die hard, and to this day there is
as much sentiment as business in the keeping of cattle, as
the primitive pastoralist views the matter.

Just as in Homeric times the women must utter formal
lamentations when the ox was slain, though it was but for
a feast, so now in the Nile valley the Dinka indulge on
similar occasions in weeping that is not entirely conventional,
seeing that the man who has actually reared the beast can
never bring himself to eat its flesh. Partly, no doubt, this
is sheer policy on the part of the wily savage, who does not
want to annoy his animal friends more than may be necessary.
It is, for instance, taboo to seethe the kid in the mother's
milk, for the obvious reason that, if this happens, the mother

will go off her milk, and small blame to her ! But there is more in the respect displayed towards the sacred animal—for example, towards the cow in India—than a calculated cajolery. The source of so many benefits to mankind is regarded with truly devout feelings of wonder and gratitude.

Nevertheless, primitive worship is apt to express itself in a curiously negative way by means of taboos that in themselves simply remove the sacred object from the category of common things. Thus for the Todas of the Nilgiri Hills the milk of their sacred cattle is itself too sacred for ordinary use. Turned into buttermilk, however, it has been desanctified and can therefore be drunk with impunity. Thus the dairy where this process is carried out becomes a sort of temple, with the dairyman as the priest—an office exacting strict chastity and all manner of other abstinences from worldly joys and pursuits. The dairy is made with a partition, on one side of which are ranged the sacred vessels containing the unconverted milk, while on the other are the profane vessels ready to receive the product from which the taboo has been lifted. Especially sacred is the pot, kept buried in that holy of holies, the cattle-pen, in which is stored the ferment in the form of some sour milk, kept over from the last brew, that is needed to start the process of coagulation. No doubt hygienic considerations support religious in providing a more wholesome nutriment in a climate in which milk does not keep. At the same time, no biochemical view of the transaction would yield the accompanying emotions of reverence and thankfulness which by treating it as a beneficent mystery these simple souls are able to experience.

In similar vein one might discourse at length on the mystic devices associated with fowling, or again with fishing. Among the Malays, for instance, whose acquaintance with the higher religions is often but skin-deep, the pigeon-wizard invites the birds to his noose by requesting them in King Solomon's name, that word of power, to enter the king's audience hall and don his breast ornaments and armlets. This is quite in the style of the spider inviting the fly into his parlour. Or there is the story of another wizard who, when his comrades had long toiled at the nets and caught nothing, flung leaves into the water and by singing over them turned them into fishes, so that a miraculous haul forthwith resulted. Negative means will likewise lead to the same ends, as, for instance, the use of a taboo language, so that the game may not understand what is on foot, as doubtless they would do if their ordinary names were mentioned. Moreover, for a Malay to speak of a Buddhist monk —though he may refer to him darkly as a " yellow robe " —would be as unlucky as it would for a Scots fisherman to talk of the minister under like conditions.

It remains to notice all too briefly the attempts of man to bring persuasion to bear on the plant life so necessary to his existence. So much, indeed, has been written on vegetation rites that Andrew Lang once protested at the undue prominence given to the subject by what he called " the Covent Garden school of mythology." The rice in India, the maize in Mexico, the wheat among ourselves are literally " mothers " to those whom they feed, and are honoured and besought accordingly. Right back to the beginnings of agriculture in Europe and perhaps beyond can be traced the cult of a great female divinity, with large breasts and swelling hips, whose fecundity enriches the earth and all that therein is. Now, the various ceremonies connected with the stimulation of the soil have their darker side, and much blood, not only animal but human, has been shed as a spiritual fertiliser. Here, however, it will suffice to glance at some of the more familiar and kindly aspects of these age-long and world-wide practices as they survive along our own countryside to-day.

Thus in the Midlands Plough Monday, the first Monday in January after Twelfth Day, was within living memory, and perhaps is still, celebrated by carrying round a gaily decorated plough from house to house, those who drew it along being called the Plough Bullocks and wearing bunches of corn in their hats. Skipping alongside and rattling a money-box for contributions was Bessy, a man dressed up as an old woman, and formerly furnished with a bullock's tail that hung down behind. All the party must jump about, the higher the better, no doubt because the corn would be sympathetically induced to grow tall in like proportion. Bessy in particular must exert himself, or rather herself; for this may well have been the Corn Maiden in person.

All over the British Isles, however, the most typical representative of the Corn Spirit was the " maiden," or " queen," or " baby," that is, doll, consisting in the last sheaf to be cut, which was thereupon dressed up as a human figure and placed over the kitchen door or the chimney hob to ward off witches and bring good luck the winter through. When the last stalks of the standing crop had to be severed, the reapers would slash at it in turn blindfolded, or else would throw their sickles at it from a distance ; the reason perhaps being that thus they distributed and concealed the responsibility for dealing the fatal blow.

In the Highlands they make up for it later in the day by drinking each their glass of whisky to the sheaf, crying, " Here's to the Maiden." We hear, too, of local taboos, such as that in one place the sheaf must not be allowed to touch the ground, and that in another no girl of questionable virtue must lay a finger on it. The like need of chastity in dealing with a holy thing is probably to be discerned in the not uncommon requirement that the cutting of the last sheaf should be left to some innocent child.

Thus side by side with the harvest festival celebrated in church linger pagan customs instinct with a natural religion that is the common heritage of man from days when his material resources were slender, and his greatest asset was an unconquerable hope.

10. RELIGION AND TRADE

Welfare and wealth are not convertible terms for the savage, who, like the rest of us, seeks for the one, but, unlike most of us, does not greatly trouble about the other. Moral values weigh more with him than material values. In other words, his relations with his neighbours, whether human, sub-human, or superhuman, are of more importance to him than his standard of comfort, which is low. Thus it would be quite unscientific to try to interpret his history in the light of our political economy. To treat the savage as fundamentally a wealth-seeking animal amounts to sheer anachronism—a projection of the present into the past.

Indeed, in imposing our civilization on people of the primitive, that is, old-fashioned, type of culture and mental outlook we are apt to forget that an improvement of the physical conditions in which they live will not compensate in their eyes for an utter disorganization of the spiritual conditions. A little sympathy will go farther with them than all that money can buy. No doubt as such unsophisticated folk learn our ways they will become more like us ;

who, having more fully exploited the environment, have our satisfactions bound up with the use and abuse of an infinite variety of possessions.

Here, however, we have to do with the beginnings of the exchange of commodities among people of simple wants, and the first question to be asked is why they should hanker after alien goods at all. It might seem the natural thing for them to do ; but the truth is that their instinct is rather to avoid the stranger and his belongings, so that a long process of fraternisation is required to enable them to overcome this aversion.

For the primeval society is self-sufficing. It keeps to itself and fends for itself, seeking peace and safety in more or less complete isolation. Of course, in proportion as its organization is at all complex, there will be lesser circles—clans, families, and so on—within the wider circle of tribal intimacy. On the whole, however, whether closely or distantly associated, all are comrades, who prefer to give and take among themselves rather than to buy and sell.

To speak of communism in this connexion, as is often done, is to go too far, because, though some things such as food are shareable and for the most part are actually shared, other things such as ornaments and weapons are privately owned, being, as it were, vehicles and extensions of the separate personality of the owner. So much is this the case, that, when a man dies, those articles that bear the stamp of his individuality and, as the Australian natives say, " smell of him," are often buried with his body as unfit to be used by others. Infected as they are with his ghostly presence, they would be as dangerous for the rest as though they carried a curse. Conversely, if a man willingly parts with one of these objects so redolent of his very being, he is virtually making a second self of the receiver. From this point of view to exchange gifts is almost to exchange souls —very much as is likewise implied by the rite of blood-brotherhood. So bride and bridegroom exchange rings to symbolise their spiritual union.

It is on such lines, then, mystic rather than economic, that

most of the formal give-and-take, as contrasted with the
casual sharing, must be interpreted in reference to societies
of the simplest type. Gifts are primarily tokens of regard,
whether towards equals, when they imply affection, or
towards superiors, such as parents, parents-in-law, tribal
elders and so on, when they express respect and the desire
to honour. After all, there is nothing strange in such an
attitude towards the employment of property ; for we ourselves would feel it indecent if the spirit of the shop were to
intrude into our relations with our nearest and dearest.
It is only when individual gain supplants communal good
will as the be-all and end-all of social intercourse that the
economic man, a characteristic product of civilization,
appears on the world's scene.

Now, however much it may wander about in search of its
daily food, the primitive society normally possesses a territory
in the sense that it recognizes boundaries beyond which lies
the country of the stranger, who is likewise for the most
part the enemy. Even that mild-natured and feeble cave-
dweller, the Vedda of central Ceylon, is wont to define his
frontiers by cutting a mark representing a man with a drawn
bow on the trunks of prominent trees along the limit—a
plain hint to the next group as to what would occur should
they be disposed to intrude in search of game or of the
delectable combs of the rock bee. It may be added in regard
to such property marks in general that they are essentially
warnings, and tend to embody a threat of more than merely
physical penalties in store for those who do not mind their
own business.

In the Solomon Islands, for instance, it is customary to
taboo a fishing-ground or a garden or a particular tree,
say a coconut or breadfruit, by placing on it a mark, for
example, a dead branch, to signify that a protective incantation has been duly sung over it. Trespassers are threatened
with running sores, distended stomach, head on one side,
and so on. Such precautions are all the more necessary
because, in addition to light-fingered gentry, there are
wizards, and, again, ghosts that need to be kept at a safe

M   177

distance. In short, supernatural man-traps and spring-guns in use all over the primitive world go to prove how deeply the spirit of landlordism is implanted in the human bosom.

For the matter of that, however, the same holds of the poaching tendency, and it is by no means natural to man to refrain from helping himself at the cost of another, if that other does not belong to his own circle and crowd. Thus we hear of more than one Australian tribe that used to send expeditions to raid the land of the stranger when they needed red ochre with which to paint their bodies, or the leaves of a certain bush called "pitcheri" that were chewed for their narcotic properties. A regular war-party was organized, and it would fight its way through hostile country for a distance of two or three hundred miles, returning home with marvellous tales of adventure among monstrous peoples who grew toes on their heels. Only picked men could join the muster. They must carefully arm and paint themselves for the part, they must never halt too long by the way, or fail to set a watch at night ; in a word, it was a sheer act of invasion, on a par with those forays across the border that occurred in the good old days when Scots and English perpetually coveted each other's cattle.

Nevertheless, paradoxical as it may sound, mutual aggression may be an avenue to mutually profitable intercourse. Though there is nothing to be said for rapine in itself, unless possibly that it serves as a check on over-population, yet in the course of actually "knocking up against" each other a certain grudging respect may be engendered, together with a desire to acquire whatever attributes and possessions are supposed to make the other side powerful and lucky. The same line of thought that leads a warrior to partake of his enemy's heart may take the mitigated form of promoting interchange of complimentary presents. Peace-making, in fact, is usually accomplished in this way, and the underlying idea is that a real bond is formed by thus participating in one another's intimate belongings. It is, for instance, a regular method of terminating a blood-feud among the Australian natives.

Nor is it a very far cry from these occasional forgatherings to periodic meetings between groups who would seem to enjoy quarrelling almost as much as fraternizing, but ingeniously make use of these contacts so that, ill will being exhausted, good will has a chance of coming to its own. There results something that is a cross between a tournament and a fair. Recriminations are hurled at each other, weapons are brandished, and perhaps a regulated combat ensues, or more often a culprit stands up passively to a shower of spears which he dodges until his assailants have blown off their steam. Thereupon all is peace and festivity, and a general swapping of effects—essentially a generous form of commerce, since it amounts to an interchange of olive branches—becomes the order of the day. Thus, historically, trade is of the same progeny as war—a younger sister with a kindlier face. As a native put it to Howitt, its object is " to make friends."

Yet making friends is none too easy among wild folk who tend to be as suspicious of one another's intentions as two strange dogs. After all, there is some reason for this attitude. For even under more civilized conditions the distinction between commerce and piracy sometimes remains a fine one. Thus in Homeric times a trading vessel was apt to snatch up a woman or two and be over the horizon by daybreak ; while in our day the natives of a Pacific island have often been kept guessing whether barter or " black-birding " was the true business of the white man's visiting schooner. Hence there is a justifiable caution at the back of the institution known as " the silent trade." I deposit a stone implement from my tribal quarry in a clearing one evening, and next day it is gone and an opossum skin has taken its place. Of course it is a little awkward if what I really wanted was a bone nose-plug or a hardwood spear ; but by taking some things and leaving others each of us may gradually come to form an idea of the other's needs. The only merit of the system, in fact, is that the mystery attaching to the foreign article is so enhanced that it acquires special value as a luck-bringer. Thus in Ashanti " suman " are fetishes or charms that consist of knotted porcupines' tails, of snail shells, of beads, and so forth. As a native informant explained to Captain Rattray : " Suman come from the ' mmotia ' (fairies), by whom they were first made and from whom they are still obtained. You place ten cowries on a rock, go away ; on your return you find your cowries gone, having been replaced by a suman."

On the whole, however, it is simpler if the two parties can learn to trust each other, and in Australia they have thought out an ingenious, if by no means obvious, way of bringing this about. A child's umbilical cord is preserved and tied up in the middle of a bunch of feathers, being then known as a " kalduke." Two fathers of different groups exchange these objects, thus causing their sons to become " ngiangiampe," that is, taboo to one another. When they have grown up, these individuals become the agents through whom all tribal barter is carried out. The only trouble is that they are barred by the very sacredness of the tie between them from ever speaking together ; so that their dealings can only be conducted through third parties. Thus from the merely practical point of view this mode of creating spiritual union has its drawbacks.

But the savage puts communion before commerce, and in his preoccupation with the mystical aspect of the transaction pays little heed to its material consequences. By this symbolical exchange of life-tokens he makes palpable to all the notion of a friendship void of all offence, because exempt from all the incidents and accidents of ordinary intercourse. It is no mere diplomatic gesture, but a solemn act of religion which thus, as it were, weds the two tribes together in the persons of their representatives, so as thus to bring a blessing on all their subsequent relations. For us, however, who have lost this way of looking at life, it is very hard to see a connexion between brotherhood and business, or to imagine how the spirit of Christmas could be extended from Christmas presents to Christmas bills.

That such an institution can remain difficult for us to
appreciate even though established on a grander scale is
proved by the parallel case of the "kula" or so-called
trading-voyage of the Trobriand Islanders, as described by
Dr. Malinowski in his fascinating but bewildering account
of these "Argonauts of the Pacific." Just as the Greek
heroes who sailed in the Argo—"the Swift"—were after
the Golden Fleece for its magical value, so these Melanesian
mariners, who in their outrigger canoes maintain inter-
tribal communications between islands scattered in a wide
ring, are primarily identifying the main interest of their
lives with a non-utilitarian purpose. On every island and
in every village there is a limited number of persons who are
"in the kula" for life, and it is their privilege to participate in
a constant exchange of white-shell armlets and red-shell necklaces. These move round and round in opposite directions,
and bind these men of different groups in honourable
partnerships, which impose obligations of mutual service in
proportion as they consecrate and adorn the social tie.
The more important of these ornaments have individual
names, like our Crown jewels, and the owner for the time
being feels uplifted, as does also the friend to whom he
passes on the precious object in due course, always as a gift
and never on the basis of a *quid pro quo* bargain.

On the whole, concurrently with this romantic business,
there goes on a good deal of minor interchange of useful
articles, and no doubt the absence of haggling that as a matter
of ceremonial decency must accompany the transference
of the symbolic gifts tends to spread to these subordinate
affairs that come nearer to ordinary marketing. Whereas,
then, the modern economist would appraise the "kula" as
no more than an indirect stimulus to sea-trading, for the
Trobriander it is in and for itself a consuming passion
whereby he is moved to take a positive pride in not receiving
more than he can return.

We have only to glance, however, to another part of Melanesia
to see how commerce based on the use of a currency, together
with the beginnings of a banking system and of a taste for

the higher mathematics, can evolve out of the trading of pieces of shell of various colours to which different values, and often individual names, are attached. Mr. Armstrong in his excellent book on Rossel Island gives a very full account of the monetary system there in vogue, the complications of which turn on the fact that only particular " coins " have the precise purchasing value required for particular transactions ; so that, however rich he may be in shell-money of other denominations, a man who needs a precise coin must start a series of exchanges that go on until the required token comes to his hands. Hence there has come into existence a regular class of financial operators who, in much the same manner as a London bill broker, make a profit by lending on shorter terms than those on which they borrow, using magic to stimulate their debtors to repay promptly and their creditors to be correspondingly slow in claiming their due.

Having thus acquired the right coins a man is in a position to buy the appropriate object, say, a pig, though even so it is not a matter of conducting a private deal so much as one of initiating a communal festival. Characteristically it begins with a sort of quarrel between the groups concerned. " Not much of a pig, that ! " says the spokesman of the one side ; to which the reply of the owner is, " Eat it and see ! " Compensation being handed over in the shape of the conventional equivalent, which includes coins of so sacred a character that they must be handled with reverence and in a crouching attitude, there follows a feast which is definitely supposed to adjust the relations between the two parties.

One more illustration must suffice of this primitive tendency to subordinate the material to the spiritual benefits of the exchange of goods. The " potlatch " or distribution feast of the north-west coast of America is on the face of it a display of reckless prodigality, inasmuch as the clan that gives entertainment strips itself bare of its property in order to send away its guests loaded with complimentary gifts. Did it stop here, they would have lost everything except honour. It is incumbent on the visitors, however, unless they would be disgraced, to hold similar feasts in due course, so as to repay the favours shown them with no less prodigal interest. Yet it is surely a noble form of rivalry not to be outdone in generosity rather than in the art of snatching a profit.

So, too, then, all over the savage world the prime function of most of the ceremonial giving and taking is to betoken mutual regard and a readiness to share as is proper among friends. Very false, therefore, are the contrasts drawn by early travellers between the hospitable folk that met them at the shore with freely tendered gifts and the thieves who boarded the vessels in order to lay hands on everything that took their fancy. In no case is it likely that the savages in question did not envisage their duties as hosts and their rights as visitors as equally binding clauses of the law of nations. Prone as they are to ill will towards the stranger —for ignorance ever breeds suspicion—these simpler peoples are thorough in their reactions, and, if they turn to good will, do their best to establish a complete reciprocity. Modern commerce might well borrow from its old-world prototype something of that non-calculating, non-competitive spirit which proclaims getting to have little or no value apart from giving.

11. RELIGION AND BLOOD-REVENGE

$\mathbf{W}$e civilized folk deem ourselves to be a long way past savagery, and indeed in
many respects we are. At most we consent to remember
that our origins go back to that intermediate condition
known as barbarism. Certain institutions, however, are
common to savage and barbarian, while they likewise
persist among peoples whom it would be decidedly unwise
to insult, at any rate face to face, by suggesting that their
claim to be civilized was at all open to question.

A typical case is that of the blood feud. The obligation to
avenge the murder of a kinsman on a strict principle of
tit-for-tat is hardly less sacred in the eyes of a Corsican or an
Albanian than in those of an Australian aboriginal or an
American Indian. So universal, in fact, is the recognition of
the duty of vendetta at what may be broadly termed the
tribal stage of society that, instead of regarding it as a moral
aberration, we must, on the contrary, pronounce it a
necessary step in the evolution of the idea of social justice.

A policy of reprisals as between different sections of the same community ceases to be defensible as soon as there exists an impartial authority capable of repressing crime by duly punishing the party adjudged guilty. So long, however, as no such central control has been established, the only practicable way of preventing outrage is by repayment in kind ; for to suffer evil like to that which he has inflicted on another brings home to the stupidest ruffian that violence does not pay in the long run.

Essentially, then, the blood feud is a quasi-legal method of adjusting relations between neighbours in the unconsolidated type of human society. Wherever such loose tribalism prevails, a man distinguishes sharply between kinsmen, neighbours and strangers. Thus, on the one hand, to slay a kinsman is a deadly sin. The voice of his brother's blood cries from the ground against the murderer—that is, the very earth is blighted by his presence and will not yield its increase until he is banished the country. So in Greek legend Orestes, the matricide, must flee the land pursued by the Furies who personify the mother's curse ; nor would the foreigner harbour him, all distraught as he was and clean out of his senses, until he had purified himself. This at length he did, not merely by shaving his head—a stock way of getting rid of impurity—but by biting off a finger. Such a sacrifice caused the Furies, who had before appeared black to him, to turn white ; for thus to make a blood covenant with a ghost, since it was his mother's blood that coursed in his own veins, was bound to convert the avenger into an ally.

As for human as contrasted with superhuman retribution, nothing as a rule is done by the injured kin, the punishment being quite severe enough if left to the ghost, which is almost to say to the guilty man's conscience. We hear, however, from Australia of a horrible method of redressing trespass against the laws of the blood, which is, however, perhaps to be interpreted not as a chastisement, but rather as an act of mercy. Among the tribes of the Wotjo group the tote-mites, having slain their offending brother, chopped the body into pieces and reabsorbed it into the totem—in simpler language, ate it ! Even if kindly meant, this procedure would no doubt serve as a deterrent, since there clearly were physical drawbacks to this mode of re-rentry into spiritual communion with the kin.

On the other hand, to slay a stranger is no sin at all, but merely an incident of war ; for the savage has his full share of that race prejudice which expresses itself in hatred of the alien and his unreasonable ways. Just as the ancient Greeks spoke contemptuously of non-Greeks as " barbarians," so primitive folk are wont to refer to themselves as " the men," and to the rest of the world by some more or less opprobrious term suggesting that they are riff-raff and scum of the earth.

Even in relation to the outer enemy, however, manslaying is restrained by certain qualms of conscience which array themselves for the savage imagination in the shape of ghostly pursuers. The warrior who in fair fight has struck a blow for home and country is not exempt from these terrors, unless he disarms his foe by a post-mortem reconciliation which as likely as not may take the form of eating him, or at least tasting his blood. When a Nandi of East Africa has slain a man of another tribe, he must be careful to wash the blood off his spear and drink it up, or else he is liable to go mad. The same custom is found away in the West of Africa among the Ibo, who think that unless this is done the man-slayer will run amok among his own people. For the matter of that, it is stated that in the Italy of to-day the Calabrian who has used his dagger in a quarrel will lick the blade in the hope of going scot-free. It looks, in fact, as if man was by nature peacefully inclined, and disposed to associate homicide, justifiable or otherwise, with a sort of temporary insanity, a fit of berserk frenzy apt to pass all bounds unless curbed by ceremonies of atonement. These alone can bring the war-intoxicated man back to his normal and human frame of mind.

We come at length to that intermediate class, the neighbours who are less than kinsfolk, but more than utter strangers. The loosest aggregate of clans shows the makings of a society ; for, although a central government is lacking and there is little cohesion except when the threat of war from outside brings the related groups temporarily into line, even so they speak one language ; while, again, they have common customs and, in particular, a common religion. In such circumstances war to the knife is inconceivable, though a certain amount of mutual bickering will be likely to result from a rivalry hardly to be deemed unhealthy, since it serves to keep one and all up to the mark. Moreover, the constituent clans or kins will almost invariably be of that one-sided type which bars marriage with either the mother's or the father's side of the family altogether. Thus each depends on his neighbour to this important and indeed vital extent, that a man has to look to another kin for a wife.

Exogamy, as this practice of marrying outside the clan is called, regularly coexists with the blood feud. Such vendetta amounts to nothing else than a mode of squabbling with one's " in-laws," whether actual or possible. So well recognized, indeed, is this fact, that in Corsica the favourite way of putting a stop to a spirited series of blood-lettings between families is to arrange a wedding ; though it is lucky if the ceremony comes off without a fresh outbreak of hostilities between the two sets of guests. It is no wonder that in rude societies woman so often fulfils the gracious function of a peace-maker ; for she may well be seeking to prevent her husband and her brother from doing their best to destroy one another.

On the other side of the account, however, must be set the fact that by their extravagant lamentations over the dead the women do much to incite the men to thoughts of vengeance ; and one might even suppose that, if primitive funerals were accompanied by less violence of emotion, there might never have arisen the belief that every death is due to someone's evil magic, which is the curse of aboriginal Australia. By egging one another on to a pitch of hysterical excitement that seeks discharge in any direction, they readily swing round from sorrow to rage and seek a sedative in some precipitate and pernicious murder.

Fortunately, however, for savage humanity, which could scarcely survive if this theory of a death for a death were strictly carried out, there exists in the primitive breast an infinite capacity for make-believe. Hence, so long as the ritual of custom is enacted with dramatic thoroughness, the will is taken for the deed, and passion exhausts itself on the symbol. Thus among the Arunta of Central Australia no one dies, however old or decrepit he may be, without someone else being proved responsible for it by divination on the part of the medicine man ; who, for instance, if an animal burrows near the grave, decides that the culprit is to be sought in the corresponding direction. It is thereupon incumbent on the relatives to form an avenging party or, as they call it, to " go kurdaitcha." The man who does this takes a medicine man with him, and they carry charms to make them strong, invulnerable, invisible and so on. When they overtake their victim they are supposed to spear him magically so that he falls, but afterwards comes back to life without knowing what has happened to him ; though in the end he assuredly sickens and dies in consequence. Before starting on such an expedition both of them must have gone through a painful operation involving the dislocation of a toe. Now it is obvious that this make-believe slaying is pure myth ; and yet many natives can proudly display such dislocated toes, and their self-importance is much enhanced by the mystery that surrounds them in the eyes of those who have not " gone kurdaitcha."

On the other hand Sir Baldwin Spencer actually witnessed the despatch of a large and well-armed avenging party who, eschewing the magical procedure that called for broken toes, went off after a death in camp to destroy someone living 130 miles away on the pretext that he must have " gone kurdaitcha " against their kinsman. They were quite successful ; for, although the man they wanted had got wind of it and had discreetly cleared off, they managed to spear his father. No doubt in due course the old man's group would pay a return visit to their neighbours to take toll of them. "In this way," writes Spencer, "year after year an endless kind of vendetta is maintained among these tribes, though, fortunately, it sometimes happens that there is more noise than bloodshed."

For noisy and tumultuous as any meetings of wild folk must be who have unsettled grievances against each other, custom prescribes a nascently legal, because a regulated, procedure for bringing the dispute to an end. One way is to substitute for the collective bear-fight a duel between selected champions. Each man presents his head in turn for the other to strike with a club—their pates are fortunately very hard—until blood flows, anger evaporates, and honour is satisfied. Another way is for the accused man to stand up to a shower of boomerangs and spears, dodging them as best he can, without attempting any counter-attack, until his assailants have worked off their angry feelings. Once decided, the matter cannot be reopened. For example, if the two groups fraternize so far as to exchange gifts—which at that rude level of society is much the same thing as to trade—all causes of quarrel between them are wiped out. This latter rule foreshadows what in later times becomes a regular substitute for the vendetta, namely composition for homicide by the payment of a blood-fine. Most of the early law of our own country is concerned with the assessment of "wergelt" or blood-money. Nay, relatively savage peoples, such as American Indians, indulged in elaborate estimates in kind, if not in money, of the price of a life as distributed among the members of the guilty group. No doubt in hot blood they would usually kill. If, however, the murderer could escape for the time being, gentler counsels would prevail, and the relations might hope to buy him off, though always at a cost that made manslaughter an expensive luxury.

To round off this account of the blood feud from a legal point of view, it is to be noted that the transition from the private to the public type of criminal law is gradual, so that many intermediate forms occur. Though the central authority, whether represented by a tribal council or by a single chief, may be fairly strong, it may hesitate to intervene in quarrels between subordinate groups except in a more or less advisory capacity. Thus a judge is hardly more than an arbitrator if, after delivering sentence, he leaves it to the injured party to exact punishment or compensation as best it can. On the other hand, an ambitious ruler will soon discover that fines can be diverted from their original function of making up for private losses, and turned into profits whereby the government can be suitably rewarded for maintaining public order.

Again, since the primitive overlord tends to be king and priest in one, it will naturally fall to him to preside over those religious accompaniments of early law, such as notably the ordeal and the oath, which impart to its decisions something of the force of divine decrees. Indeed, whereas the private blood-feud might be said to treat homicide as a crime, it is reserved for the public trial, which as it were but interprets and carries out a judgment of heaven, to impress on it the character of a deadly sin. Hence, as compared with the more loosely organized societies in which as often as not fining quickly brings fighting to an end, the centrally governed state is from the first severe, not to say harsh, in its punishments, and the nascent citizen is hammered into shape with a heavy hand. As all along it involved a curse to slay within the kin, so henceforth in respect to the whole society, being as it were an enlarged brotherhood, it becomes abominable to shed blood.

It remains to say something about the moral aspect of the blood feud. A savage who prosecutes a vendetta has no notion that he is indulging in unseemly hate or lust for blood. On the contrary, over and above the plain duty of keeping the fighting strength of his group level with that of its rivals, he perceives in the act of avenging his dead an obligation dictated by affection, by honour, and by conscience itself.

Thus a native at work on an Australian station, and supposed to be more or less tamed, lost his wife. With the logic of his people he instantly concluded that she was the victim of evil magic. Reproved by the white man for his foolishness, he went about his duties for a while, steadily growing thinner, however, and obviously pining. Indeed, he asserted that the ghost of his wife kept appearing to him and reproving him for his slackness. At length he stole off, and was no more seen for a month. On his return he was sleek and happy, so that there could be little doubt as to what had happened in the meantime.

Or, again, though we move upwards from this low grade of culture to the verge of civilization, we find little change in the ethical outlook. In Serbian the word for taking blood in a feud is " osvetiti," which also means to hallow or consecrate. Revenge, in short, is a religious rite. Miss Durham quotes a Montenegrin author of 1860, who writes thus : " Osveta (vengeance) is something born in a man. The Montenegrin would rather die than live shamefully. The ' osvetnik ' tracks the ' krvnik ' (man who has taken blood). He cannot work, he cannot sleep. . . . In truth he must take blood, otherwise he has no place and no honour among the Montenegrins."

Though a mother knows she will probably lose her son if she sends him out, at the age of fifteen or so, to slay or be slain for his father's sake, she does not scruple to produce from her dower chest the blood-stained piece of the dead man's coat which, as all believe, becomes moist so that " pervlon gjak," the blood boils. Nay, the soul of the murdered one cannot rest until blood has been taken for him. Wherefore the Albanian, despite the refusal of his religion to countenance such barbarism, declares " that he would rather be in hell with his honour clean than in heaven with it black."

When, therefore, from our civilized standpoint we condemn, and rightly condemn, the whole spirit of the old-world practice of the blood feud, we must remember that, if we are no longer justified in taking the law into our own hands, it is simply and solely because the law of the sovereign state has taken over from private hands the function of suppressing outrage by the use of force. Whenever the central authority is not up to its duties, it becomes incumbent on the citizen in his individual capacity to take up arms on behalf of justice and right ; and on these lines a defence of lynch law is possible, if, though only if, the state is impotent to keep the peace. Otherwise, granted complete impunity, crime will stalk through the land, and the mad dog will spread its poison throughout the pack.

For the rest, of all the problems of morals none is more difficult than how to reconcile justice with mercy—to administer punishment where it is needed without disregarding the healing power of forgiveness. Whereas savagery scarcely looks beyond the needs of the group, we civilized folk have gradually come to study the needs of the individual ; and, the more we do so, the greater is our unwillingness to judge any man bad beyond all saving. At least, then, we have taught ourselves to hate the crime rather than the criminal, who thus when he suffers the extreme penalty is hardly more outcast than victim—a vicarious sacrifice whereby we remind ourselves that to go back on society is to go back on life itself.

12. WAR-CHARMS AND LOVE-CHARMS

Making love and making war have more in common than might at first sight appear —at any rate, in their primitive forms. For normally the savage is exogamous—that is, marries out of his own social group into another which has more or less different customs and entirely distinct obligations. If a quarrel between the two clans arises, as over a matter of blood spilt or a death imputed to sorcery, a man will have to join the avenging party that is going to attack his wife's people, and may actually be obliged to slay someone who is her brother or father, whether by blood or by tribal convention. In any case, short of such occasional hostilities, his intercourse with her relatives will be extremely formal, complicated as it must be by various avoidances, of which the taboo on any sort of intimacy with the mother-in-law is the stock example. Thus such a marriage might not unfairly be compared to what happens among ourselves when an Englishman weds a foreigner who is likewise of a different religion ; for to belong to a separate totem almost comes to that. The only mitigating circumstance is that everyone has to do the same, the tribe as a whole forming a sort of inter-marrying union such as causes the component elements in the long run to get used to one another's ways. To go right outside the tribe for a wife is unusual ; and though it can be done by peaceful arrangement—not, however, without much wagging of elderly beards over the question how a matrilineal system is going to be adapted to a patrilineal one, and so on—it is more often a simple case of making off with the captive of one's victorious spear.

Moreover, in addition to this more or less casual mating with a stranger-group a man is necessarily brought likewise into contact with a stranger-sex. Crawley may have made rather too much of what he terms the " sex taboo " when he tries to show that the main object of marriage ceremonies is to neutralize this mystic horror instinctively felt by each sex for the other. Certain it is, however, that sex solidarity under conditions of primitive culture tends to be very great ; and it is only a step, emotionally, from want of sympathy to active fear and dislike. The men and the women have each their own industrial pursuits, their own initiations and other mysteries. No wonder, then, that they are divided in soul, even if the imperious urge of the life-force drives them into union as concerns the body.

It needed, then, a long process of development in order to vindicate the cause of peace, whether domestic or political ; nor are the results thereby attained even at this time of day altogether satisfactory. There can be little doubt that under civilization moral has lagged behind material progress. Nevertheless, as contrasted with the rude beginnings of culture, both the family and society at large involve to-day in far greater measure an effective recognition of the claims of individuality, alike personal and ethnic ; and on such a basis only can mutual toleration, and as its outcome equal love, be solidly established. Here, however, no attempt can be made to review this evolution of charity in its entirety, or even to estimate the part—the very important part—which has been played therein by religion. Distinguishing religion from magic by the motive underlying the one or the other, we may class as magical those influences which on the whole have been reactionary and bad—attempts to exploit super-natural means for unworthy ends as dictated by anti-social feelings such as lust and hate.

In applying this test, however, we are faced by the diffi-culty that moral standards for the purpose of history must be treated as relative to the stage of culture under considera-tion. Thus, as regards marriage, for instance, a savage moralist who saw no harm in wife-lending, but rather a duty incumbent on the hospitable, might nevertheless condemn wife-stealing as the blackest of sins. Or, again, in the matter of war, in all ages the patriot has been apt to vocifer-ate hymns of hate against the public enemy ; while, so long as the vendetta ranks as a more or less legitimate method of redress, it will seem perfectly proper to vow vengeance on one's next-door neighbour. Difficult, however, as it may be to draw an exact line in all cases, the magical practices that will be here noticed are in general those which, not only for us onlookers but for the people concerned, count as hardly respectable, and, however dimly, are felt to militate against the natural kindliness of human relations.

To illustrate how utterly different are the ways of savagery from ours, and at the same time how diversified and con-fused are the institutions of the so-called simple life, let us glance at the love-making of the Kurnai, a tribe belonging to the most primitive stratum distinguishable among that very primitive race, the Australian aborigines. To begin with, it must be mentioned, even if it is not strictly relevant to the present topic, that each sex has its own peculiar totem or guardian animal, the " woman's sister " being Djiitgun, the superb warbler, and the " man's brother " Yiirung, the emu-wren. Such an arrangement made flirtation easy. A young man, meeting a comely maiden, merely had to call out " Djiitgun," and she would reply " Yiirung," perhaps following it up with a question that amounted to a proposal, " What does the Yiirung eat ? " ;

after which, if he coyly responded " He eats kangaroo," or whatever it might be, they both would laugh, and the outcome might well be a match. The lady's form of proposal, by the way, is usually interpreted as an offer to act as housekeeper ; but it has been suggested that the real point is to ascertain whether his food taboo differs from her own, so that there should be no totemic bar to marriage between them.

It may be, however, that the eligible bachelors are for some reason backward, whereupon the women are moved to take up the matter and stir up love by way of strife. So they sally forth and slay an emu-wren or two, and casually exhibit the little corpses to the young men. They, in their indignation, seize sticks, the girls do the same, and heads are broken. Moreover, next day the indignant males go forth in their turn and kill some superb warblers, the women's " sisters." An even worse fight follows, so that the wounds are sore for a week. Nevertheless, after having thus broken the ice, the young people become decidedly more aware of each other, and individual flirtations and proposals of the kind already described are sure to follow.

At this point, however, might occur considerable difficulty in winning the consent of the girl's father to the plan so spiritedly conceived by the lovers. Negotiations, involving possibly some exchange of women, would prove tedious, and, one may guess, might mean that a young man must wait until his elders and betters had been served. In any case, too, violent means of obtaining a wife were not unknown among the Kurnai ; for we hear of one clan raiding another and despoiling it of all its womenfolk at one fell swoop, a regular battle resulting. In such a moral atmosphere, then, it is no wonder that our pair of lovers should think of a solution of the practical problem that recalls the method of Gretna Green, with a magician to take the place of the obliging blacksmith. This worthy was known as the Bunjil-yenjin, " professor of love-spells," and his speciality was to cause elopements, whether the object of desire was maid or wife. One native whose spouse was thus induced to desert him could repeat the very charm that was her undoing, one so powerful that it " made the women run in all directions " when they heard it. The translation is as follows : " Roll up the twine, little sweetheart. I go first to the hollow before you." The reference to the twine suggests some accompanying act of mimic snaring or tying a knot. Another spell of no less efficacy is of even obscurer import : " Why did Yiirung cut off his beard ? Djiitgun sleeps in her mother's camp."

There is no attempt made to conceal the fact that the magician is at work, and some female friend is sure to inform the girl that the words of power have been launched against her. Moreover, such is their force that her parents feel compelled for the time being to cover themselves up as if asleep. As soon as the couple are well away, however, etiquette demands that an avenging party should start on their trail, and, if they are caught, the girl might be speared in the leg, while the young man must undergo an ordeal by combat with the relatives. These would even sometimes employ a rival magician to direct their pursuit. If, however, the pair got away, and turned up later on with a baby, all was forgiven. Altogether, the proceeding was irregular, but winked at ; and magic was made the excuse for getting round the law—never a reputable proceeding, but still at times convenient.

Again, another Australian tribe, the Arunta, who also indulge in runaway marriages, which, we are told, are more irregular in appearance than in reality, have charms wherewith to wile a woman away, and it is found in practice that there is no such opportunity for this as a dance, when the handsome young men are prancing about in all their bravery, with the women looking admiringly on. One has only to possess a shell ornament called a " lonka-lonka," which flashes in the dancer's waist-belt, more especially if he has sung over it beforehand to charm the lightning into it. When a woman sees it glitter " all at once her internal organs shake with emotion." A headband that has been sung over, called " chilara," is also most effective, or, again, the sounding of an "ulpirra" or conch. The woman thereupon becomes, in native parlance, "much infatuated," and, even if another man's wife, is ready to bolt with her charmer, although at considerable risk of sudden death ahead for either or both.

Passing on from these very savage folk whose love affairs are so unintelligible to us, we may note that even relatively civilized people such as the Malays do not disdain to employ a medicine man to decoy the object, or, perhaps one should say, the victim, of their passion. The charms have a strong Arabic cast, and mostly consist in requests to jinns or angels to bring off the affair, their most striking feature being the ideal of beauty therein revealed. A desirable male is one who, in addition to a voice like that of the Prophet David, and a countenance like that of the Prophet Joseph, has a tongue like a magic serpent, teeth like a herd of (black) elephants, and lips like a procession of ants. Female loveliness, on the other hand, involves eyebrows that resemble pictured clouds and arched over like a fighting-cock's spur ; a nose like an opening jasmine bud, a neck with a triple row of dimples, a head like a bird's egg, and lips like the fissure of a pomegranate. It may be added that, whenever such personal attractions were likely to wane, the medicine man could provide spells that would enable his client to restore them by "living backwards."

In ancient Greece a Thracian witch, whose spells were potent enough to draw down the moon, might well make light of such a simple matter as recovering a strayed lover. All that was necessary would be to catch a wryneck and spread-eagle it (if the Irishism be allowed) to a wheel, which was spun round while the wretched bird, still alive, emitted its magic cry, "Iu, Iu." This penetrated to the very heart of the swain. The modern Greek maiden prefers a bat for the purpose. She roasts it to cinders on a fire of sticks gathered by a witch at midnight, and buries the remains where cross-roads meet. It is also just as well to burn incense over the ashes for forty days, before the powder is slipped into the drink of the reluctant youth. It may be doubted, however, whether any young man is worth that amount of trouble.

In the island of Guernsey the same result may be obtained more simply by baking a cake of flour, salt and soot, eating half of it, and wrapping the other half in a garter taken from the left leg, after which it is placed under the pillow ; whereupon at midnight the double of the adored one will stand by the bedside. That this is really devil's work, how-ever, is proved by the story that when this had been tried on a young officer the wraith left a real sword behind it ; and later on, when the marriage had taken place, the bridegroom came upon the sword in turning out the contents of a chest, and was instantly moved to plunge it into the bosom of his lady. The moral is that love can be trusted to work spells of itself without the aid of magic, which in motive is always selfish and therefore cruel. To make a waxen image, paint a heart thereon, and transfix it with a pin is alike potent to enamour or to slay, as if in either case there was a like brutality of purpose behind the act.

Turning to the subject of war magic, one must presumably class under the separate head of religion appeals to a God of Battles which, even when they have ceased to be blood-thirsty in tone, are apt to remain almost as one-sided as they were in the days when one tribal deity frankly went forth to overthrow another. Verging on magic, however, are those more or less private explosions of suspicion and hate that are exemplified by an Australian avenging-party. Here the medicine man assists from first to last ; for to begin with he uses divination to detect the guilty person, and later on may be required to accompany the expedition and do impossible things—in which nevertheless the natives firmly believe—such as mending up the victim after he has been speared, so that he does not know that he has been attacked, though he languishes away and dies none the less. For some obscure reason everyone, including the medicine man, must have previously undergone the ordeal of dislocating a toe ; and, if their subsequent doings remain enwrapped in clouds of myth, certain it is that such dislocated toes

abound and are treated as badges of honour. Nay, a woman
will occasionally go on such an avenging mission when
someone of her own sex has offended against the proprieties
by not cutting herself properly at a funeral. She carries a
charmed stick which is supposed, when thrown from behind,
to enter her enemy's body in small pieces, which only a most
expert medicine man could extract by sucking.

Usually deemed magical because "sympathetic," or, one
might almost say, telepathic, in their imputed action are
those ceremonial observances which, all over the savage
world, are incumbent on wives whose husbands are away
collecting heads or otherwise engaged in warlike pursuits.
The almost universal requirement that the women should
observe strict chastity might seem to have more than a
purely ritual reason behind it, though in theory it is a means
of keeping the vigour of the absent husband unimpaired.
On the other hand his womenfolk are free to dance their fill,
as he will thereby be rendered the more active. The dance
may itself be symbolic, as when African wives, with an
ancient sorceress at their head, ply brushes in order that the
men may sweep away their foes, or, more gracefully, fans
are agitated in the Kei Islands so that the enemy's bullets
may be waved aside. Sometimes an opposite effect is
ascribed to the same prescription ; for in Malaya the
husband's sleeping-mat must be rolled up, presumably that
the suggestion of an interloper should be excluded, whereas
in Celebes this very thing is taboo, because it would imply
that it would be long before the man could be back. For the
rest, it is obvious to the primitive mind that, if the wife
dozes during the day, her lord will be drowsy, if she oils her
hair, he will slip, and so on.

Meanwhile, the combatant likewise must arm himself with
magical aids, both offensive and defensive. Thus a fighting
man's ghost is known in Melanesia as a " keramo," or ghost
of killing. Hence the relic of such a man, such as a tooth
or a tuft of his hair, is worn in a bag hung round the warrior's
neck for luck. Indeed, the relic itself is called the " keramo,"
and in its name he curses the enemy, crying " So-and-so

eats thee ! " Another Malay spell combines offence with defence, as thus, " I apply the charm of the Line called the Swollen Corpse," where the reference is to a line drawn with the foot, beyond which the foe is unable to pass by force of this dread formula, which likewise indicates his fate after reaching it.

Finally, the well-worn device of the wax figure is not to be despised in this connexion. Thus an Egyptian king defeated his foes at sea by placing wax models of the ships and men in a bowl of water and duly consigning them by his spells to perdition. So, too, an Arab writer of the thirteenth century accounts for the victories of Alexander the Great by declaring that Aristotle had furnished him with wax images of his enemies nailed face downwards in a box that must never leave his hand. Such a tale, however, sheds more light on Arab than on Greek mentality ; for it needed but a breath of Greek rationalism to blow all such nonsense to the four winds of heaven.

If the civilized victim of a festal season calls in a well-groomed gentleman with a good bedside manner to prescribe the necessary pill, we speak of the doctor and his science. If, however, the sufferer is a savage who, having partaken too freely of a rather tough kangaroo, commissions a personage with wild eyes and a bone through his nose to scare the devil from out of his stomach, we refer less respectfully to the medicine man and his magic. Yet for the anthropologist these are but different stages of one and the same evolutionary process. At the same time, it is only fair to note that a radical change has occurred not only in the method employed in each case, but in the theory behind the method. The doctor assumes that he is working with a set of natural causes; whereas the medicine man on the whole relies on what we should be inclined to describe as supernatural causes, though he might not have an exactly equivalent term in his own vocabulary.

Thus it is interesting to consider the history of our phrase

" medicine man." It goes back to the early part of the nineteenth century, when Americans coined the word " medicine " to express anything and everything that had to do with the religion of their Indians ; to whom they ascribed medicine dances, medicine feasts, medicine songs, and so on, while animals, stones, arrows, badges, lodges were equally entitled to the prefix " medicine " if they entered in any way into the native scheme of sacred rites. Catlin, the intrepid explorer-artist, who in the 'forties ventured far out into the Western prairies in search of his models, puts the matter in a nutshell thus : " The word medicine means mystery and nothing else."

There exist, indeed, in many of the Indian languages expressions more or less answering to this notion of mystery ; though the precise meaning would mostly seem to be that of " mysterious power." The vaguer term " power " is better suited to render the sense here than " spirit " would be, because to the primitive mind it remains undecided how far life and personality attach to the occult agency ; more especially since it is supposed capable of being passed on from one person or thing to another person or thing.

Now in one important respect our word " medicine " fails to do justice to the original conception, because the latter likewise extends to what we should call poison. In other words, whatever is supernaturally potent can cure or can kill according as it is used with good or with evil intent. Either, then, we must speak of good medicine and bad medicine when white and black magic, religion and sorcery are severally involved ; or we must reserve " medicine " and " medicine man " for those occasions in which a beneficent purpose can be taken for granted in the thing done and in the doer of it.

Preferring, then, this second course, let us give the subject of witchcraft the go-by. The witch, after all, is a rather unreal person. Savages, no doubt, stand perpetually in great fear of him, and put down all the ills of the flesh to his diabolic machinations. From time to time, moreover, some unfortunate wretch is accused and duly convicted of the abominable sin of bedeviling his neighbours, and dies some dreadful death in order that the tribal nerves may be quieted. But for obvious reasons a real witch, did he exist, could not afford to come out into the open. Still less might any association of witches hold their sabbaths with impunity for long, however secret they tried to keep them. The chances are, then, that with savages, just as with ourselves in the days of the notorious witch-trials that disgraced Europe in the sixteenth and seventeenth centuries, the victims of popular panic immensely outnumber those authentic criminals who in some hole-and-corner way practise the black art for reasons of spite or greed.

On the other hand, the so-called witch-doctor is no witch at all, but something a good deal nearer to a priest. True, he may be little better than a hedge-priest ; for there are grades of respectability in primitive religion, and those who serve the greater gods of the tribe are superior in social status, and doubtless in moral outlook as well, to the keepers of local shrines and petty talismans—in a word, to miracle-mongers of the baser sort. Generally speaking, however, whatever the society concerned approves of, or, at any rate, tolerates, as religious is more or less good in their eyes, and depends for its validity on genuine faith in which all share alike, both the priest and the people.

From the standpoint of the civilized onlooker, critical and unsympathetic, who cannot participate in this believing attitude, they may seem one and all the dupes of what he ranks as mere superstition. But that this priest-like wonder-worker is a conscious fraud—though this is exactly what the vulgar are apt to think about those who profess any other form of religion than their own—is a psychological absurdity. The faith-healer who does not have implicit faith in his own power of healing will never produce the like conviction in his patient. No dishonest man ever made " big medicine " ; for the laws of mind forbid it. Even the successful quack believes in his own quackery.

Perhaps the best way in which we can learn to appreciate his state of mind is to watch the medicine man in the making rather than as already trained for and engaged in his profession. It is to be noted at once that it is not every man's job, but one that proves attractive only to individuals of a special and comparatively rare disposition. Thus we are expressly told about one Australian tribe numbering about 1500 souls that it could show no more than one or two medicine men to each of its four divisions. Of these few it is hardly too much to say that each has answered to a " call," finding in himself a capacity for supernormal experience and thereupon of his own initiative following his peculiar bent. One of them, for instance, spoke of his boyhood to Howitt thus : " I used to see things that my mother could not see. When out with her I would say, ' What is out there like men walking ? ' She used to say, ' Child, there is nothing.' These were the ghosts which I began to see."

Now the ordinary initiation through which every youth must pass would involve fasting and other austerities likely to develop latent talent of this type. But the candidate for orders must go on to face a further initiation. Our information is unfortunately meagre as to the precise nature of these even more searching tests, but all native testimony agrees in stating that the power thus acquired comes by a kind of " revelation "—whether it proceeds from ghosts, or from other spirits similar to our fairies, or from the supreme god of the tribal mysteries. In any case the power is no mere matter of technical skill, but ranks as something super-humanly imparted—a divine gift.

Of course human agents help to supplement spirit-action in the course of such proceedings, being, however, men who have already graduated as past masters. We hear of one of these, for instance, transmitting virtue to the novice by the laying on of hands, though he was careful at the same time to paint on the latter's back the sacred design representing the very spirit from which he himself had received his own power. Moreover, it is incumbent on such a candidate for the sacred office of a healer to undergo the ordeal of sleeping in front of a cave where the spirits are reputed to dwell ; and the official explanation of the fact that he afterwards appears with a hole bored right through his tongue is that the spirits do this with an invisible spear. Now self-induced stigmata have been alleged to occur among holy men ; but it is simpler to assume that the work of the spirits is in such a case carried out by human substitutes. And yet in fairness to the savage one must be chary of imputing conscious fraud, since to impersonate a spirit amounts to being one for the primitive actor, who throws himself into the part sanctioned by tradition with the utter self-abandon of an unquestioning faith.

As for the dazed youth who purchases admission into the transcendental world at such a price, he is bound for the rest of his life to associate the exercise of his mystic vocation with a certain strictness of behaviour. If he ceases to be holy, his gift of healing is withdrawn. There are well authenticated instances of doctors who, when the weakness of the flesh had caused them to break their taboos—as by eating fat, or, again, when one of them took to the strong drink of the white man—felt themselves become impotent, and had the decency to retire from practice forthwith. Now it may well be that alike in Australia and in the rest of the primitive world all members of the profession are not equally conscientious : and indeed it is sometimes reported that the native doctors themselves are moved to use forcible means to suppress the quack. But one is bound to judge any class of men by its best and not by its worst specimens. Given a universal belief that disease is a miraculous visitation only to be met by equally miraculous counteraction, it is natural to regard oneself with complete conviction as the vehicle of a supernatural power or grace, such as will work wonders so long as a certain ritual process is carefully followed. Healing, in short, is an inspiration, not a science, in the eyes of those simple folk who are at the mental stage known to psychologists as that of " primitive credulity " ; though " faith " would perhaps be a less question-begging name for it.

Turning now to the methods employed by the medicine-man, we may take it that in general he is carrying out something analogous to a rite of exorcism. It follows that, even when his treatment resembles the modern doctor's attempts to combat disease in a physical way—as, for instance, if massage or a hot fomentation is used—the primitive practice is based on a non-physical theory, and amounts rather to the casting out of a devil by making it too uncomfortable for him to haunt the patient any longer. We must therefore be careful not to put a literal interpretation on acts which are intended to be purely symbolic. Thus the Australian leech will sometimes make play to extract the mystic evil from the sick man's body in the visible shape of a quartz crystal, which some ill-disposed person is supposed to have shot into him by the force of his curses. Or, conversely, he will effect a cure by apparently producing a similar crystal out of his own body and causing it to disappear into the body of his patient.

Now, physically considered—and this is the way in which the ordinary European bystander regards it—this is mere make-believe. That is indeed exactly what it is ; but to make the patient believe that the evil has gone out of him, or that a counteracting virtue has gone in, is the point of the whole performance. As a matter of fact, the savage mind has to struggle hard against a materialistic view. Thus, during the medicine man's initiation not only is his body thoroughly scarified with such crystals in order to fill him full of healing-power, but he actually drinks a few small ones in water. No wonder that a native account, translated literally, says : " Always crystals in the inside of the doctor are—in the hand, bones, calves, head, nails." It is worth noting, too, that, although the manipulation of real and palpable pieces of quartz is calculated to reach the patient's imagination by way of his senses, this is not invariably done. For the doctor may content himself with making passes that are supposed to project the crystals invisibly, or, as they say, " like the wind " ; though even the wind metaphor tends to be taken literally, so that the sick man is apt to hear a whistling as the life-giving emanation is wafted from the doctor's body into his. Just as in a case of exorcism by means of holy water it is not the water, but the holiness that ought to be deemed efficacious, so at this lower level of confused thinking there is a considerable mental effort required to distinguish between outward sign and inward grace. Hence, for instance, we find side by side in the vocabulary of a Queensland tribe two names for the medicine-man, one being, " crystals-many," and the other " the man full of life." To make an end of the subject of these magic stones—which, by the way, on account of the prismatic colours which they display are thought to come from the Rainbow—there is good evidence that the doctors know how to use them so as to produce hypnotic effects, a fact almost sufficient in itself to invest them with a spiritual function and meaning.

One might proceed to explain on similar lines many another device of the medicine-man which on the face of it looks like a piece of clumsy conjuring, and is therefore usually derided by the superior white man, who prides himself on seeing through the trick. One end of a string being tied to the patient, the medicine-man puts the other end into his mouth, rubs his gums until they bleed, and thereupon spits out the trouble for all to see. Is the doctor necessarily a humbug, and his audience a set of deluded fools, because the ritual takes the form of a rather telling bit of drama ? If speaking took the place of acting, and the disease-demon were bidden to come forth, the symbolism might from our point of view seem more refined, but its essential character would not be altered. Indeed, the primitive doctor, though undoubtedly he relies on pantomime for his most striking effects, is also wont to deal with spiritual evils by word of mouth.

If we insist on dubbing his whole procedure " magic," we shall have to class these verbal formulæ as " spells." Now admittedly these often resemble commands rather than petitions ; though it may be questioned whether this is not the proper tone to take with devils. Often, however, there is prayer, or something like it, by implication, as when the doctor chants endlessly, " Show—belly—to the Moon," with great stress on the last word, to call down beneficent influence from that mystic quarter. Explicit prayer, however, hardly occurs in Australia.

Among more advanced peoples, however, it is quite common to address polite requests to the disease-spirit to go away. One may even seek to propitiate the Small-pox by calling him " Grandfather," though it might be rash to contsrue this as a token of genuine affection. Nay, such a fell being may be given the status of a god or goddess, when perhaps sheer terror, which is surely the natural attitude, is transformed into a kind of reverence towards one whose chastisements may be but the just penalty of human sin. In any case, however it may rank, whether as influence, power, spirit, demon or god, the object of the doctor's activities is nothing belonging to this world, nothing that can be classed as animal, vegetable, or mineral, but is of that other world which is beyond natural law and human reckoning ; so that a certain arbitrariness may be expected of it, whether a policy of bluffing or of humouring be followed. Meanwhile experience has taught the savage that a most effective way of facing unknown risks is to hope strongly ; and the key to the medicine-man's operations is that they are attempts to induce this condition of mind in his patients.

Magic, then, does not seem quite the word to cover dealings with the mystic side of life when their intention is so well-meaning. It is sometimes said that anything is magic which implies " My will be done," rather than " Thy will be done," the sentiment underlying true religion. Now perhaps the magic of hate does imply the former of these attitudes, since the very act of hating tends to isolate a man from the rest of his kind. But it is pretty certain that the doctor who has been initiated by the spirits does not think of himself as controlling those spirits. On the contrary, it is just the other way about. He is not controller, but controlled. They possess him, and work their will through him as a mere medium. Even if his voice issues commands, these are their commands, not his. All those phenomena which intrigue our modern spiritualists are equally familiar to primitive folk, and in the department of practice, if not of theory, they may even have more to teach than to learn.

Thus one very paradoxical way of getting the better of a disease-demon is to submit to his control. Major Tremearne described very vividly the bori dances of certain Hausa folk now settled in Tripoli, in which the dancer obtains power over the small-pox, or whatever the plague may be, by letting the dread spirit enter into him, the result being that it becomes harmless, since, on the principle of " the hair of the dog that bit you," it cannot harm itself.

According to Major Tremearne nearly all the bori are disease-demons, but the exact functions of many of them are not defined. Speaking generally, it may be said that, the more important the bori, the more definite is the disease which he inflicts ; but in many instances the spirit's power is indicated by the number of illnesses which come from him, and, at any rate, the young bori can be responsible only for rashes of some kind or other and for sore eyes. Ignorance of the special powers of a particular bori is often due to the absence of a proper mount, or dancer ; for in time the bori would tend to be forgotten. No Mai-Bori, that is, initiate into the mysteries of the bori cult, except one who had learned the proper actions, would dare, or be allowed, to invite a bori to ride him ; for any mistake in the ritual would bring the bori into contempt, and the dancer and the whole audience would be liable to his displeasure. The idea is that the bori are summoned in order that they may torment those people who are prepared for them, and thus be more likely to leave the rest of the community in peace.

One might almost treat the practice as a dim anticipation of curing by inoculation, were it not that in the primitive view the transaction is not on the physical plane at all. It is altogether in another direction that we must look for our explanatory clue, namely, among those religious experiences bordering on ecstasy in which resignation to a superior will begets an access of borrowed strength.

Be the ultimate reason what it may, the savage, too, has

discovered that out of weakness can come forth strength, out of submission of will renewed will-power. In his own way he has explored secrets of the soul which we with our greater interest in the control of our material resources may be inclined to neglect. Enough, at all events, has been said to show why the anthropologist cannot roundly and uncritically rate the medicine-man as an impostor, and his so-called magic as a transparent fraud. Relatively to the prevailing state of society and of knowledge, he is a worthy who does his best by his clients ; nor is his best by any means wholly bad.

14. TABOO

The traveller who wanders about the trim suburbs of Honolulu is constantly faced by a notice-board set up at the entrace of some tempting garden and bearing on it the single word KAPU. Like the "tapu" of New Zealand or the "tambu" of Melanesia, this is but a variant of "tabu," the form of Oceanic speech used in the island of Tonga, whence Captain Cook brought it back to adorn the English language under the more homely spelling of "taboo." Now the American landowner who addresses this curt notice to native trespassers has really no business to use so strong an expression ; for he is virtually saying, "This is holy ground !" He ought to mean, in fact, that persons who are rash enough to intrude will not only be duly prosecuted and punished in this world, but will likewise be eternally damned in the next. Indeed, in the olden days, if a Hawaiian chief who ruled by divine right had hoisted such a danger-signal, his subjects to a man would have credited him with the power of blasting the offender alike in body and in soul. In modern eyes, however, neither the institution of private property nor the law itself can claim a sacredness so absolute as to involve the direct authority of Heaven.

In its ultimate derivation taboo seems to mean " marked off," just as our word sacred may have originally meant " cut off " ; that which is so demarcated and set apart being the supernatural in all its manifestations, whether good or bad. In the Pacific region the word which comes nearest to our " natural," and is regularly opposed to taboo, is " noa," that is common or ordinary. The savage observes that both men and things have their more or less regular habits of behaviour ; and so long as they keep to these, he for his part can hope to accommodate his habits to theirs. Within this limit of the habitual, then, life is pretty safe, though no doubt correspondingly humdrum. Beyond it, on the contrary, all is uncertain ; and, when the unknown has to be met, the first rule is " Beware ! "

Taboo, then, is just this preliminary counsel of prudence. Discretion is the better part of primitive religion ; though it can ripen into a genuine reverence as the conviction grows that behind the mystery lies an influence, or being, that is not only incalculably powerful but also incalculably good. The original reaction, however, towards the uncanny leaves it doubtful whether harm or help is likely to be forthcoming ; and, if Man were all coward—and in any case he is decidedly not all hero—one might have expected him to give a wide berth to whatever baffles common sense. But, apart from the fact that troubles are apt to come uninvited, we are one and all compounded of cross-impulses, such as fear and hope, disquiet and curiosity. Shout at a flock of sheep and they will start running, but presently they will stop to gaze round in search of the cause of the disturbance. A more intelligent animal exhibits even more clearly a like interplay of tendencies making severally for repulsion and attraction. Thus there is the well-known story of the ape at the Zoo which, when required to retire for the night, refused to leave the outer cage, and sat on a lofty perch gibbering defiance. It was only necessary, however, for the keeper to enter, and, gazing down a drain-pipe, to pretend a moment later to start back in horror. Thereupon the ape descended slowly and took a cautious peep for himself, only to bolt instantly with a cry of terror into the safe retreat provided by the inner compartment. So too, then, there is for the primitive mind an intriguing as well as a purely alarming side to any experience of the unfamiliar ; so that the taboo sign by no means implies an absolute ban on contact, but simply enjoins a wise circumspection in the manner of approach. Thus more especially does the notice apply to idle meddlers. There is no admission to the world of the sacred except on business. An unlimited liability attaches to this kind of venture, and, if any profit is to come of it, the perpetual watchword must be " Attention ! " What the Roman called " profane " persons—as we should say, worldlings—had better keep at a respectful distance.

One series of experiences to which every savage is normally subject will serve excellently as illustrations of the general nature of taboo. There are at least four occasions in the course of his earthly career, namely, when he is born, when at puberty he is initiated or made into a man, when he marries, and, finally, when he departs this life, which are deemed so critical as to demand special precautions on the part of all concerned. They are, in a word, sacramental seasons, and a corresponding mood and behaviour are needed if the slippery ground is to be safely crossed on this pilgrim's progress. Society in its religious capacity sympathises with the human weakness associated with birth and death, or again with the shyness incidental to the assumption of the responsibilities of manhood and marriage. So a scheme of rites is devised to permit a certain retirement from worldly affairs. It is in its way a reciprocal relation that is thus set up, because, just as the subject of the experience, together with those in immediate touch with him, is held to be in a spiritually delicate condition requiring freedom from disturbance, so ordinary folk engaged in the common round do not want to be troubled by the presence of such mystic invalids in the midst of their daily bustle. It is thus to their mutual satisfaction that they should keep away from each other. What the profane man finds good company is bad company for the sacred man, and vice versa.

First of all, then, a man has to be born, and in so doing is a trouble not only to himself but to his parents, and more especially to his mother. The latter as soon as she knows that a baby is coming must practise various avoidances to keep herself and her precious burden from harm—not mere physical harm, but the thousand immeasurable risks that lurk in the unknown. Indeed, her views as to the nature of conception may be so pre-scientific that she may ascribe it to the influence of some animal or plant. Thus a woman on Mota in the New Hebrides found an animal in her loincloth and tried to carry it back in her hands to show it to the village, but lo ! when she opened her hands it had disappeared. Clearly this was no common beast, and it was incumbent on her coming child never to eat such meat for the rest of its life ; for in a similar case an eel-child all unwittingly tasted eel in a mixed grill and promptly went mad.

Again, what the mother eats may affect the looks of the child, so that a Kaffir mother who does not want her baby to have too large an under-lip must never touch the underlip of a pig. In general, pregnancy is a period of seclusion ; and, though many of the accompanying restrictions are to our mind unnecessary and even absurd, there is sound experience at the back of the system as a whole. Indeed, there is practically much to be said for trying to meet and satisfy the nervous imaginings that are so regular a symptom of that condition ; while as a matter of scientific fact it is not impossible that the mental impressions of the mother may in certain circumstances be reflected in her offspring, as folk-belief has always held.

Passing on to the actual birth, the cleansing of the child is treated not merely as a physical but likewise as a spiritual necessity, and it is duly baptized or, it may be, half choked by a prolonged fumigation. When the infant cries in the

smoke, the Kaffir mother calls out, "There goes the wizard";
just as at a christening the nurse may be heard to rejoice in
baby's lusty protest as a sign that the devil is being expelled.
Again, it may be thought expedient to perform a ceremony
to purify the mother's milk before it can be safely drunk.
Too many indeed to mention are the taboos which the
savage mother must observe until some rite of purification
restores her to her workaday status.

Nor is the father immune from similar restraints, but on
the contrary is often subjected to a confinement almost as
strict as his wife's, at any rate on the non-physical side.
Much cheap fun has been made of the custom known as the
" couvade," from the French word " couver," to go broody
in the manner of a sitting hen. Thus Strabo, the ancient
geographer, wrote that in Spain when a child is born it is
the man who is brought to bed. As a matter of fact the
Brazilian native does take to his hammock on such
occasions, because if up and about he would be bound to
do something unsuitable—to use a cutting tool, for instance
—that might jeopardise the welfare of the child. Papa,
mamma and baby must all lie low until the domestic crisis
is over. It is holiday for them all, in the primitive sense of
the word, that implies a fast rather than a feast.

Next, puberty is eminently a time of taboo. Our educational
experts have but recently come to realize that this organic
crisis is likewise a moral one ; so that it is highly inadvisable
to force the young during the period in which they are
storing energy to be expended later with renewed vigour
on a plane of fuller life and greater responsibility. The
novice who undergoes initiation must retire to the wilderness, where austerities of all kinds are imposed on him by his
elders as a preparation for the stern duties of manhood.
In the meantime, however, he is regarded as neither man
nor boy, but an intermediate being, a sort of embryo on its
way to be born into a new condition ; and the metaphor of
a new life is variously enacted by ritual means. Nay, so
whole-heartedly does the savage enter into the spirit of his
drama that, when the probation is over, and the boy turned

man—the convert, as one might say—comes home in triumph to don the accoutrements of a warrior, his own mother is not supposed to recognize him. He is as strangely different as a butterfly that has emerged from the chrysalis ; for his soul has found wings. His new personality is sundered from his old by a sort of hiatus in the sense-bound experience of ordinary existence, during which he has been withdrawn into the subliminal depths where the deferred instincts of sex and parenthood are sprouting as inevitably as the down on the young man's chin. It is quite a fitting symbol that on his return the regenerated one should simulate complete loss of memory as regards common things and even words. On the other hand, the spiritual experiences of his stage of transition abide with him ; so that, for example, if fasting has brought him a vision of the animal destined to be his mystic guardian and helper—his " genius," as the Romans said—he must make it his private taboo for the rest of his days, refraining from injuring or eating it, or even from mentioning it casually.

Marriage, the third of these typical sacraments, needs a long chapter to itself. Suffice it to say that honeymooning is not a modern invention, since as far back as we can penetrate into the mists of human history it has been plain to all that, without some gradual progress of mutual adjustment involving a liberation from the stale round of the day's work, the couple must miss that slow-dawning sense of a larger life in store for them, and theirs, which is the true consecration of their union. Sometimes among savages the period between betrothal and marriage is stressed as taboo, sometimes that between actual mating and the birth of the first child. In any case, bride and bridegroom are by means of all sorts of ritual prescriptions marked off as involved in the holy state of matrimony—holy for themselves, but for their neighbours unclean in the ceremonial sense that they had better keep clear of whatever is thus set apart.

This double aspect of the matter will explain, by the by, those contradictory features so often noticeable in marriage ceremonies ; for it may chance that either of the happy (or unhappy) pair is now touched for luck, and now pelted to avert misfortune. It is a mistake to try to put too precise a meaning on such customs. They arise in the first instance out of a thoroughly confused state of mind on the part of all concerned, in which not only allied feelings such as fear and respect are mingled, but there is a downright pivoting between opposite tendencies, fear and hope, tears and laughter. A sort of hysteria, an upsetting of the normal equilibrium, is bound to occur when the clash of interests, traditions, temperaments caused by a marriage has to be overcome by a sort of humouring process.

Not only are the two sexes designed by nature to be different though complementary, so that a certain reserve in their dealings with each other, a certain concession to each other's privacy, is desirable ; but the two families have to learn to tolerate more or less discordant ways and become friendly. Curiously enough, savage society often seems to contemplate the other possibility of their becoming too friendly, though this is but one side of the relation of mutual aloofness typified by the polite distance maintained between the son-in-law and his wife's mother. It is felt in such a case that too near is too far. Social beings though we essentially are, we individually need to guard the approaches to our own soul-world ; and never is this more urgently required than in time of emotional stress, when the only remedy is quiet as a means of self-concentration.

Finally, death might seem an end rather than a period of transition were it not that, from the earliest cavemen onwards, the forefathers of our race have had imagination enough to reckon on a future life, and have organized their funeral rites on that comforting supposition. On the other hand, death always comes as a sudden break with present existence ; and, just as those who are left behind need time in which to turn round and face the new situation, so the deceased him-self might be expected to linger on the threshold of that other world. So closely indeed is the attitude of the mourners reflected in the mood attributed to the dead man that the conventional term of mourning usually sets the limit to his spell of hesitation. Thus for a year, it may be, the shade still haunts its former place of abode, so that a second and concluding funeral may be required to send it away, whether to dwell with the ancestors for ever or else to return in due course as a reincarnating spirit. During the same interval the bereaved relatives must faithfully wear the trappings of affliction and otherwise mark themselves off as disinclined to mix with their fellows. Nay, the ghost would resent any carelessness in this respect ; and an Australian widow, who, being dark-skinned, wore white gypsum as her mourning garb, explained that this was to catch her late husband's eye and to assure him of her sorrow at his loss.

A day arrives, however, when, however regretfully, the living must cease to look back, in order once more to attend to the mundane affairs so long interrupted ; while in turn the widow herself may wish to remarry, or perhaps as the wife of her deceased husband's brother may go on to bear children that continue the former's name. As for the ghost, it, too, must abandon its half-way standing and, after a tender parting, go for good.

If indeed it " walks " after that, there is something wrong with it. Either the funeral rites were scamped and those responsible are in for trouble ; or it has its own reasons for being unquiet, as, for instance, after a suicide, when a stake at the cross-roads might be needed to " lay " it. Nay, all unburied dead, like unbaptized infants, pine in the homeless middle state ; whereas those who have been duly seen off on their passage—sometimes with the fare actually provided in cash—are sure of a welcome among the shades, with or without the possibility of an eventual re-birth, when the cycle is resumed and the re-invigorated spirit by way of the taboos of infancy prepares once more for life and its troubles.

Here, then, are four crises, of which two in modern eyes are natural events, namely birth and death, and two social mechanisms, namely education and marriage, that the savage prefers to class as mysteries, perhaps not without reason. Though they primarily concern the individual, they likewise affect his intimates, so that one and all are simultaneously provided with an off-time for spiritual rest and refreshment. It is hard to say whether they take it, or it is conceded by the rest, who, after all, have no use for those whose heart is for the time being not in their daily work. Instead of merely smiling, then, at the eccentricities of the taboo system with its tendency to perpetuate the accidental phantasies and phobias to which we all give way when temporarily off our balance, one should note rather the essential needs of the human spirit to which these rough and ready Sabbath ordinances minister with marked success ; and can test each for himself the value of the primitive view that there are difficult times when it is best to have it out with oneself in quiet and alone.

15. TOTEMS

Of all the generalisations of anthropology none is more certain than that the ancestors of all existing peoples have passed through the hunting stage. True, such hunting must be taken to include mere gathering, since one tribe may hunt chiefly for elephants and another for acorns. One and all, however, who thus go straight to nature for their nutriment are alike in being, not food-raisers, but mere food-collectors. Life is a perpetual game of hide-and-seek as played between them and the other denizens, animal and vegetable, of their immediate environment. No wonder, then, that animal-lore and plant-lore together form the very sum of their philosophy. Further, with but rudimentary appliances to help them in the chase, and confronted by beasts as fierce and almost as cunning as themselves, men are less conscious of being, as they are to-day, veritable lords of creation ; but humbly deem themselves very much on a moral level with the brutes—part and parcel, in fact, of the democracy of the wild, a welter of perpetual faction-fights, in which foes are many, and friends few but correspondingly valuable.

To illustrate this primitive attitude of equality towards what we now think of as the lower creatures, one has only to consider the world-wide belief that birds and beasts talk their own languages—nay, that if a man is clever enough to understand what they say he can pick up a marvellous amount of curious and useful information. One still hears people remark, " A little bird whispered me that secret," and in such a case it would be only fair to inquire how the knowledge of bird-speech was acquired—whether it was by picking up a magic ring, for example, or by eating a serpent's liver, or simply by putting fern-seed into one's shoes, all of these being well-tried specifics. Thus, in the story of the Boy who became Pope, he annoys his father by learning nothing at school except the languages of the dogs, the frogs and the birds ; but this turns out to be quite enough education for him to qualify as a social climber. As for the hunting savage, it is told, for instance, by Sir Everard im Thurn, in his " Indians of Guiana," how round the camp-fire at night his native companions would spin the most impossible yarns about their conversations and communications with the furred and feathered tribes of the forest ; not only commanding implicit credence from others, but evidently tending to believe it all themselves. For the matter of that, many of us are prepared to hobnob with a favourite dog or horse almost as if he were a Christian.

It is, then, against such a general background of a naïve sense of fellowship with all living nature that the special facts should be viewed which anthropologists class together rather loosely under the head of totemism. Current definitions of the term are none too satisfactory, but for our present purpose it may suffice to say that totemism is that complex of beliefs and institutions which is based on the mystic self-identification of a human group or individual with some non-human natural kind. The word comes from America, and occurs first, in the form " totam " and " totamism," in an account of his travels written in 1791 by an Indian interpreter named Long. Among the Chippewas, a large tribe inhabiting the region of the Great Lakes and belonging to the Algonquin-speaking stock, *ototeman* translated literally meant "his-own-kin," and thus stood for a man's clan—in this case the group that counted relationship through the mother's, not the father's, side of the family. The word then found its way into American literature ; so that, for instance, Longfellow sings in "Hiawatha" :

> From what old ancestral totem,
> Be it Eagle, Bear or Beaver,
> They descended, this we know not.

It was, however, the British anthropologist, J. F. McLennan, who in 1869 published a paper on "The Worship of Animals and Plants" which raised the term "totemism" to the status of a principle applicable to the whole of the primitive world. To-day one has only to consult the four immense volumes of Sir J. G. Frazer's "Totemism and Exogamy" to realize that, in view of the endless ramifications of the subject, one is almost entitled to speak of a totemistic stage of human society. Only a few hunting peoples show no traces of such a custom, and, with perhaps the single exception of the Eskimo, they are very backward in respect to all the arts of life, and indeed may well have degenerated. Meanwhile, though the provincial forms of totemism, as they may be called, are very various, so that, for instance, Australia, North America and Africa can each boast a type of its own, there are also common features so numerous as to warrant the guess that we are in the presence of a very ancient distribution of culture from some central cradle-land, such as South-western Asia.

Be this as it may, we have not the evidence to bring home definite totemism to palæolithic man in Europe, however much we may suspect it in view of the preoccupation of the cave-artists with the figures of animals, including those of animal-headed men, who are presumably meant for masked dancers. We know at least that in Australia the aboriginals divide up among themselves the responsibility of persuading the different animal kinds by means of symbolic ritual to be fruitful and multiply, so that, as described in the chapter on "Religion and the Means of Life," the men of the kangaroo totem do their best to stimulate the kangaroos, the emu men bring their influence to bear on the emus, and so on. Hence it is not unlikely that the impressive sorcerer of the cave of Les Trois Frères, with his reindeer mask, may have styled himself a reindeer, and that there were likewise human cave-bears or bisons, whose business it was to keep in mystic touch with their animal counterparts and incite them to do their duty by mankind.

On the other hand, hunting-rites intended to control or conciliate the game are not necessarily totemistic, but form a wider class within which falls the particular variety involving the totemite's belief that his totem is of one nature with himself, and must therefore be in sympathy with his wishes and needs.

Indeed, however preferable it may be to stick to known facts, it is tempting to indulge in speculation concerning the mysterious origins of a scheme of thought and behaviour so widespread and so alien to our civilized notions. As plausible a theory as any is that given at the outset by McLennan who supposed totemism to arise out of a system of naming individuals or groups after different kinds of animals or plants ; all the further developments being due to the fact that in savage philosophy the name is taken as equivalent to the thing named, which in this case happens to include both the borrower of the name and its original owner. Indeed, if a savage came thus to see in his namesake animal or plant a sort of second self or double, he could hardly fail to feel a respect for it that might easily ripen into awe. Moreover, he might reasonably expect help from such a source as being an extension of his own personality ; while, contrariwise, to injure it would amount to suicide by proxy. Of course, there lurks in the background the further question why such animal or plant names should, in the first instance, have been bestowed on individuals and groups. We might be clearer on this point if we knew whether the individual name or the group name came first. Thus an individual name, if given at birth, might have something to do with the imagined cause of the birth in question ; for primitive notions on this subject are apt to be entirely non-physiological. Thus in the New Hebrides conception is attributed to some animal or even some fruit with which the mother has been in accidental contact ; it being thereafter incumbent on her child to abstain religiously from food of a like nature. On the other hand, a group might come to be known as the bear-folk because they chiefly hunted bears and were well posted in all the lore conducive to that purpose. All this, however, is mere conjecture, and, even when the savage professes to explain how he came by his totem, we may be sure that he is following our example and simply drawing on his imagination.

Here, for instance, is a tale told by the Tsimshians of British Columbia. Once upon a time a man met a bear in the forest. Instead of attacking him, the bear took him to its home and there it taught him many clever and useful things. So when at length the man came back his sister made him a dancing-blanket on which she painted the figure of a bear. Whereupon her children took the bear for their crest, and were ever afterwards known as the bear clan. So explains the primitive anthropologist, making up a Just-So story with a fancy that draws no line between men-folk and bear-folk, unless it be in favour of the latter.

Most typical of all, however, among the ways of accounting for these animal or vegetable family names is to invent stories of descent from supernatural ancestors of mixed traits ; so that an Australian will speak of one of his fore-runners of the great days of old now as a man-kangaroo and now as a kangaroo-man. At most the natives occasionally show themselves to be aware of a slight distinction between the two. Thus, at Undiara, the great centre of the kangaroo totem in the Arunta country, there is a water-hole, and close by a rocky ledge. The first is full of the unborn spirits, which in that state of existence have neither legs nor arms nor head, of kangaroos destined to become real men and women ; while the second contains similar unborn spirits of kangaroo animals. By cutting themselves freely and anointing the rock with their blood, the men drive out in all directions these animal spirits, so that they can enter the female kangaroos—exactly as the human spirits are believed to enter the women—and may so cause the species to multiply. Quite apart from such utilitarian purposes, it is edifying to relate stories or to hold dramatic performances in which the legendary doings of the old-time semi-bestial folk are duly celebrated ; which doings are often more wonderful than those of man or of beast as we know them now.

Thus, when a celebrated kangaroo-man called Ungutnika was wandering about in the good old days, he came upon some wild dogs, who tore him to bits and swallowed the meat, leaving, however, the skin and bones on one side. But Ungutnika was by no means done for. The skin simply covered up the bones again, and off he hopped. Three times this happened, until at last they dragged him to Undiara and ate him up all except the tail, which they buried ; and there, to prove the truth of the story, the tail may be seen to this day, turned into a stone. It may be added that during his lifetime Ungutnika was afflicted with boils, which he tore out of himself and put into the ground, where they likewise became stones. Hence, a modern kangaroo-man who wishes to score off an enemy has only to charge a toy spear with the evil magic contained in one of these stones, and he can pass on the boils merely by pointing the weapon in the other man's direction.

Or, again, to dress up as an emu ancestor is an elaborate affair involving the construction of a long, tapering head-dress of grass stalks bound round with hundreds of yards of human hair string. The performer then struts about with a swaying motion that suggests the aimless peering about of the actual bird. So, too, a couple of the eagle-hawk people make up a play by pretending to be eagle-hawks squabbling over a piece of meat.

All this mummery may seem to us rather pointless, but it is the most solemn and moving thing in the life of the native, enabling him not merely to assert his right to his own distinctive name in the sight of all, but, over and above that, to acquire that sense of increased power and worth which the glamour of ancestry invariably confers, whether one claims to be descended from a totem or from a god, from a prophet or from a simple pirate.

Indeed, the interest in ancestry may develop to such an extent that totemism virtually passes into heraldry. This has been the case on the north-west coast of America, whence come those huge totem-poles that are the glory of any ethnological museum fortunate enough to possess them. One sees the best part of a great pine tree carved fantastically so as to exhibit here the projecting beak of a monstrous raven, there a bear hugging a man, and somewhere else a killer-whale—for a clan may possess not one only, but many, of these crests or emblems, which are supposed to commemorate events in its early history. These chronicles, however, are not strictly veracious, unless, for instance, we believe the eagle family of the Haidas when they tell how they " first came out " from an ancestress with the unusual name of " Property-making-a-noise," and she took them to the Tsimshian country wearing a dogfish tattooed on her back, whence they came back bringing a beaver, and so on.

To use the total number of such crests belongs to the chief of the clan only, and a youth, if an Eagle, starts with only the eagle crest, and can add dogfish, beaver, or what not, only after he has made lavish potlatches—that is, distributions of property—to the members of the other division of the tribe. Nay, as happens elsewhere, aristocracy among these children of Nature is not far removed from plutocracy. The leading people jealously guard possession of some famous crest, and, if a chief of lower rank were to adopt it, the head of the superior family would put him to shame by giving away, or simply destroying, valuables beyond the means of the upstart, who must thereupon relinquish such expensive honours. Truly this is property-making-a-noise with a vengeance !

From heraldry it is but a step to art, and, as in mediæval

Europe, the crest may become part of the Indian warrior's bravery ; so that, for example, wooden war helmets were carved in the shape of sea-lions, while totemic animals figured on canoes, paddles, clubs, rattles, spoons, boxes and so forth. In one way, however, the American brave had the better of the European—at any rate after the latter ceased to decorate himself with woad—because he could express his social importance by means of body paint. Thus a man entitled to the devil-fish crest would get up his face in red and black to resemble the creature, the goggle eyes staring from his forehead, while its feelers squirmed over his cheeks. Again, tattooing was resorted to so that every member of a family must carry with him a permanent proof of pedigree in the shape of a frog or a humming-bird, a dragon-fly or a codfish, sometimes more or less naturalistically rendered, but often so conventionalized that no one not in the secret could possibly recognize what was meant.

Just so, too, the totemic designs of the Australians for the most part bear little or no resemblance to the real objects ; so that a series of concentric circles or a spiral on one bull-roarer will stand for a frog, while on another it means a tree. Such sacred drawings are supposed to have been handed down from the time of their remotest ancestors with certain meanings attached ; and for them it is sufficient to know those meanings, without asking themselves how they came to be acquired in the first instance.

It remains to notice very briefly the bearing of totemism on religion—a difficult subject, if only because it is not easy to define religion so as to indicate the precise point at which respect passes into the reverence which implies worship. We may certainly go so far, however, as to say that the totem is sacred. For instance, the Bechuanas of South Africa call it their " glory." If you want to know a man's clan you ask him : " What do you dance ? " He must not eat the animal in question, or wear its skin ; nor, if he can help it, would he even look upon it, lest evil befall.

There are, however, subtle differences in the degree of

regard such as we can scarcely appreciate. For instance, one group has two totems, the hartebeest and the eland, and distinguishes between them by saying : " The dance is to the hartebeest ; the veneration is for the eland." The crocodile folk call the crocodile father and master. When the chief makes an oration, his followers must cry aloud : " Oh, crocodile man ! " at every fresh point that he makes. They swear by the crocodile. If they unwittingly go near one, they must spit on the ground and say : " There is sin." The totemite is unhappy even when other people take liberties with his animal friend. Thus the porcupine clan are distressed if anyone has killed a porcupine, and, having religiously collected its quills, they spit on them and rub their eyebrows with them, lamenting thus : " They have killed our brother, our master, one of ourselves, him whom we sing."

Of course these ritual observances are largely of a negative order, so that such religion as they amount to might almost have to be described, with Reinach, as " a system of scruples." On the other hand, to dance one's totem is to enter into coummnion with it and share its *mana* or mystic power. Whether as a clansman participating in the collective luck, or as an individual acquiring a personal protector in response to a vision obtained by fasting, the totemite by identifying himself mystically with his totem feels that he is somehow enlarged in soul-power ; and this is in essence a religious feeling, and a positive one at that.

PART IV.
PRIMITIVE TECHNOLOGY

16. ARTS AND CRAFTS OF PREHISTORIC MAN

Man is the manipulative animal ; he owes his supreme position among living creatures to his handiness. As for the tool, it may be regarded generically as an extension of the hand. Handling things thus by deputy, Man, in the course of a long process of experiment, has discovered how vastly to increase the range of his manipulative activity ; so that the mechanical achievements of modern civilization seem at first sight out of all relation to their far-off humble beginnings.

Nevertheless, the study of the origins of our present immensely complex arts and crafts not only reveals the strictest continuity between them and their savage prototypes, but likewise proves that the early chapters of the romance of industry can bear equal testimony with the later to the inventive genius of our race.

Now prehistory starts from a blank. When Man first comes into view he is already in possession of a material culture implying, on the evolutionary hypothesis, a lengthy process of previous development. Probability alone can guide us in reconstructing the tool-using efforts of fossil man's unknown precursors. A few animals, and notably the apes, which in point of physical resemblance are nearest to Man, are said occasionally to fling a stone or brandish a stick when in a state of nature, and during captivity can be taught to display such accomplishments in much greater variety ; though it remains doubtful how far in the one case or in the other such acts involve any full awareness of the function of the means employed.

If, then, we credit Man or his pre-human ancestor, despite his opposable thumb and his relatively bigger brain, with an initial technical ability on a par with an ape's, we might label it the appreciative stage ; since the user of the casual stick or stone at least appreciates it as somehow satisfying to his present mood. From this basis a rising scale can be constructed consisting of successive stages which may be called the selective, the adaptive and the inventive.

Thus, first, there is advance as soon as, in place of the first thing that comes to hand, a natural object is picked out as especially suitable for a given purpose. Moreover, such a tendency must soon lead to the collection and storing of a supply of ready-made appliances of the type preferred. Next follows the adaptive stage, when it is realized that what does not wholly conform to the required pattern may nevertheless be made to do so by subjecting it to more or less modification. The artificial tool now makes its appearance ; and it is to be noticed that there arises herewith the need for tools that will help to fashion more tools—for a hammer-stone to chip a flint knife, a flint knife to carve a wooden club, and so on. Finally, the inventive stage differs form the adaptive only in the degree in which Nature's provision becomes subordinate to Man's design— in other words, comes into being at whatever point the suggestion of the form can no longer be said to be given with the matter.

Of course, Man cannot create out of nothing, and the intrinsic qualities of the material thus always, in a sense, predetermine the use to which it is put. But it is one thing to improve on a pointed stick by sharpening it further, another thing and a cleverer to think of attaching to the end a spike of stone or bone ; so that perhaps one might class all composite instruments as inventions rather than adaptations. At any rate, when for the natural operation of flinging such a sharpened stick by hand there is substituted a mechanical propulsion obtained by means of the spear-thrower or the bow, Man's constructive imagination has clearly left mere copying behind and has embarked on a course of genuine origination.

Passing on from these theoretical considerations to a survey of the actual evidence, we clearly cannot expect many recognizable traces to be left of the appreciative or the purely selective stage. If sufficiently durable, the things doubtless survive ; but how is one going to distinguish them ? At most one can hope to judge by marks of use, as contrasted with those marks of design which are exhibited by the product of adaptation or invention.

Thus, more especially if they are associated with worked implements, one can pick out the natural pebbles used as hammer-stones by the abrasions visible on their striking-surfaces ; or may, less certainly, argue that other pebbles found in a little heap near a hearth and showing signs of having been subjected to considerable heat were cooking-stones. Or, again, I have discovered in a cave habitation of pre-glacial man pieces of rock crystal, jasper and so on, of remarkable appearance, that were presumably brought there because they somehow appealed to the fancy.

Meanwhile, the supreme test of the former presence of prehistoric man—apart from his bones, which are comparatively rare—is the evidence of workmanship. Unfortunately at the very beginning of the adaptive stage the genuine artefact is exceedingly apt to be confused with the " pseudomorph," nature's counterfeit, which sometimes reproduces the appearance of human fabrication with almost irresistible plausibility. Hence the vehemence of the controversy about eoliths. Here it would not be in point to discuss the authenticity of these Pliocene or even pre-Pliocene specimens of alleged human handiwork, since the main point at issue is simply whether Man can thereby be connected with so remote a geological horizon.

The further question which arises as soon as their human source has been proved, namely, what special purposes they are intended to serve, has hardly been touched. A few of them might perhaps appear suitable for cutting, piercing, and so on, in a rough way, but for the most part they consist, as Professor Sollas once put it, of " scrapers that will not scrape, borers that will not bore, and planes that will not plane." Indeed, I once heard an ingenious defender of certain decidedly blunt and shapeless examples explain that in those far-off days Man, being a gentle vegetarian, was mostly concerned between meals with " preening " himself, and found, I suppose, a flint rubber particularly stimulating to his skin or, rather, hide.

Without prejudice, then, to the question of the value of such evidence as demonstrating the existence of Tertiary man—and J. Reid Moir's East Anglian specimens, especially those from Foxhall, have converted many eminent authorities who before were sceptics—one may say that it throws no light at all on the state of his culture. True, certain backward savages of modern times, such as the now extinct Tasmanians, have used rude implements which, with certain instructive exceptions, could hardly by simple inspection be assigned a human origin or purpose ; though we know as a fact that the natives found them highly useful for scraping spears, notching trees to make foot-holes for climbing, skinning animals, smashing marrow bones, cracking shells and what not.

But the Tasmanians had no better material at hand than a rather intractable sort of sandstone ; whereas eoliths are suspiciously associated with precisely the spots where natural flints are common. In any case, since we could not divine

Tasmanian habits from their rudimentary stone-work considered by itself, we are none the nearer to deducing the hypothetical eolith-maker's habits from his stone-work, however like the other it may seem to be.

In the Lower Palæolithic industries, pre-Chellean, Chellean and Acheulean, the typical implement is the so-called " *coup de poing* " (" boucher," knuckle-duster and hand-axe are other terms), though sites of this age in reality yield a fair variety of other forms. Meanwhile, the *coup de poing* itself varies greatly in the course of its development, the out-line being triangular, pointed, lozenge-shaped, oval or even round, while the line of the edge, at first irregular, becomes as time goes on either quite straight or gracefully sinuous. Produced out of a flint nodule or other lump of stone, by chipping both sides until a sharp working edge results at one end and a butt comfortable for grasping at the other, this fine instrument presupposes a long apprenticeship in the art of reducing crude matter to a desired shape ; such, indeed, as is scarcely reflected in our existing Palæolithic series of types, however much we draw on the eolithic series to help it out. A possible inference is that the adaptive stage was inaugurated by an age of wood, the output of which has perished. Indeed, it is in any case necessary to suppose that wooden clubs, spears and so forth were in use side by side with the *coup de poing* ; which is usually thought— though it is but a guess—to have served more as a " general utility " tool than as a weapon of offence, as the word " *coup de poing* " would suggest.

Meanwhile, whatever its function, this form of stone implement appears to have come into universal favour ; so that it would even seem that in the Acheulaen as compared with the preceding Chellean period there was greater concentration on this single type. For the rest, it has a very wide distribution in the Old World ; and, though we cannot be sure that all the specimens belong to the same far-off age, some of them at least, notably certain South African examples, are so demonstrably ancient as to suggest that the cradle-land of the industry lies somewhere outside Europe, perhaps in Africa itself.

Whether such implements were ever hafted is uncertain, and the term " hand-axe " is not used here because it begs this question. It is extremely probable, however, that the inventive stage of uniting stone and wood in one composite tool had not yet dawned for the man of the Lower Palæo-lithic—any more than it had for the Tasmanian when first discovered, unless isolation had caused him to degenerate ; instances of lost arts being fairly common among islanders.

As for the use of other material besides stone, nothing survives from this period in the way of worked bone except a most remarkable instrument found in Sussex in association with the famous Piltdown skull, and certainly very ancient, since it was made out of the fresh thigh bone of a kind of elephant that preceded the mammoth. It is about seventeen inches long, is shaped rather like the blade of a bat, has a chiselled edge, and shows a notch on one side as if a thong had been attached to it. Two bones from Taubach near Weimar, one cup-shaped, the other like a dagger, may be more doubtfully classed as artefacts of the same remote age. Doubtful also are the " figure-stones " assigned to this period by Boucher de Perthes and others ; they are nodules which, thanks to their shape, or to chipping which may or may not be by Man, bear some resemblance to animal forms.

With the Middle Palæolithic begins the cave period, and at once we are on intimate terms with Man, being able to visit him at home, where the floor, carpeted with refuse, proves his shelter to have been also both kitchen and work-shop, nay sometimes even his place of burial as well. Brutish though Neanderthal man was in physical appearance, he must have been intelligent in his own way. In one cave of this period which I excavated I found that, of about 15,000 pieces of flint, two-thirds showed marks of use, and half of those secondary chipping with some approach to design—surely a very economic use of material. In another I looked round for a good hiding place and discovered, or, rather, was shown by a sharp-witted boy, a well-screened crevice high up by the roof; this notion, it appears, had likewise occurred twenty thousand years ago to some astute individualist, since the recess concealed the finest of Mousterian " points," in fact the catch of my season.

No wonder, then, that the earliest cave-dwellers, being so sagacious, had the use of fire ; without which, indeed, their caves would have been scarcely habitable in view of the rigour of the climate and the attentions of cave-bears. How they made it is suggested by the occurrence in their caves of iron pyrites, two lumps of which would provide a strike-a-light. How they used it is abundantly proved by their hearths, with adjacent bone-middens testifying to the rich stock of available game animals and to the effectiveness of their hunting methods. What these methods were, however, may hardly be deduced from the cave remains, and we can only suppose that the fashions of the age of wood persisted largely into these times ; since their whole stone industry is suggestive of domestic and sedentary functions, such as cutting up the meat, or scraping the skins and boring holes in them for fastening as garments, or rounding off a stick with a hollow scraper used like a spokeshave.

The typical instrument, the so-called Mousterian " point," was probably a universal tool, a sort of " sailor's knife," just as was its predecessor, the *coup de poing*. The latter, by the way, though it survives as part of the normal equipment only into the earlier portion of this period, may have retained a ceremonial value—just as the Neolithic axe became the sacred thunderbolt in the eyes of a later age—to judge from the fact that the man of Le Moustier was buried with a *coup de poing* in his hand, though it was almost certainly obsolete by this time for daily use.

As for the difference in form between the *coup de poing* and the point, it is the outcome of what amounts to no less than a revolution in technique. The one is essentially a trimmed nodule, the other a trimmed flake previously disengaged by percussion from a nodule. We need not postulate a genius who suddenly introduced a novel method which no less instantly found favour with all, perhaps because it was labour-saving. During Lower Palæolithic times the flakes detached in the course of making the *coup de poing* were utilised and even trimmed.

To specialise, however, on the flake implement and fashion it out of the choicest and most substantial segment of the nodule was reserved for the Mousterian culture in its later and most typical phases. True artistry is displayed in the most symmetrical and finished points, which illustrate, as it were, the survival of the fittest among numerous experiments, and occur in the proportion of about one to a hundred flakes in a representative Mousterian site, as Rutot has observed in Belgium and I can verify from Jersey. The well-trimmed base usual in this class of implements suggests that they were meant to be grasped in the hand.

The other tools of this period, awls, scrapers, planes and so on, tend to be rough, and are hardly more than adaptations almost forced on the designer by the accidental fracture of the material. A rather characteristic artefact, the so-called disk, which, unlike the rest, is often worked on the lower as well as on the upper surface, has been interpreted as a missile for throwing or perhaps slinging ; but it is simpler to regard it as the direct successor of that round Acheulean *coup de poing* which presumably served to cut and scrape. Such a circular edge is awkward to hold, and seems to imply some sort of hafting.

Dwarf flakes, an inch to an inch-and-a-half in size, showing marks of use and sometimes secondary chipping, were very frequent in the Jersey caves ; and I may add in illustration that, on examining a little bag taken straight from the neck of a Tapiro pygmy of New Guinea and containing his greatest treasures, I found it to be full of just such sharp fragments and of nothing else. There is no need to assume that any of these smaller Mousterian flakes were arrow-heads, or that the bow had yet been invented, as there are a thousand other uses for a miniature cutting tool.

As for bone, its only known use at this time was as a sort of anvil or chipping block on which downward strokes could be made without blunting the edge of the flake. I have also found a large boulder the flattest side of which had been dented and almost hollowed out with hammering, probably in the course of breaking up marrow bones. The pounders and hammer-stones, by the way, when I compared a series of several hundred with a similar series from a Neolithic kitchen-midden, proved decidedly larger and heavier, showing that, as his bones confirm, Neanderthal man had far more power to his elbow.

Upper Palæolithic times are the golden age of prehistoric art. This would be so if we regarded simply what one might call works of pure art, namely, painting and statuary, did one not suspect a magico-religious purpose to have prevailed in most cases ; but even the articles useful for everyday life were lavishly decorated, as if these ancient cave men rivalled the Greeks of Periclean Athens in their passion for sheer beauty.

Which kind of man it was that took the lead in this development—whether it was big and handsome Cro-Magnon, or small and perhaps negroid Grimaldi, or yet another of the divergent racial types that now begin to appear—is not at present determined ; but, one and all, these " neanthropic " types can be contrasted with their " palaeanthropic " forerunners as in all essential features akin to modern races, and, in fact, no more ape-like than ourselves. Not that we need suppose the home comforts of these later cave men to have been strikingly superior to those of Neanderthal man. I could not perceive, in excavating an Aurignacian floor, that the fare was more plentiful or varied ; while the same untidiness and insensitiveness to smells were attested by the piled-up litter.

Yet that the culture was in every way more diversified and refined is the dominant impression yielded by their remains. Even the stone industry, which except with the Solutreans was perhaps something of a secondary interest, shows many a modified form : long, delicate blades, little scrapers of quaint design, limpet-shaped, parrot-beaked and so on ;

while the typical implement, the graver, the stand-by of the artist, is entirely new. In general the coarser method of percussion flaking would seem to have been largely superseded by pressure flaking, already known to the Mousterian worker.

As for the Solutrean masterpiece, the " laurel leaf" point, trimmed on both flat faces by the finest pressure strokes— so that one can only suppose objects so exquisite and fragile to have had a ceremonial rather than a utilitarian function —it seems, with other characteristic implements such as a point with a single tang, to be the work of a people who thrust themselves in for but a while between two stages in the development of the genuine cave artists of the Aurignacian-Magdalenian tradition. The latter worked on bone, ivory and horn in preference to stone, as giving greater scope to their plastic and graphic efforts.

Some useful implements made out of bone, such as needles, polishers, spear-points and harpoons—the last named exhibiting a very pretty development, the barbs growing gradually neater in shape and appearing first on one side of the shank and then on both sides—do not greatly lend themselves to decoration. Others, however, such as the so-called wand, the spear-thrower and the arrow-straightener (if such it be, and not a chief's baton, as it used to be the fashion to term it) are often richly engraved.

Nay, the certainty of touch with which a few strong lines are made to express living forms in their essential features is altogether astonishing—as if fine art could approach perfection without waiting on the rest of culture to develop correspondingly. And of course with this artistic skill there went dexterity in all branches of workmanship, as witness the minute eyes bored in some of the bone needles by means of flint awls, themselves appropriately delicate and hard to manufacture.

Again, while we are considering the utilitarian side of these cultures, let us note how inventiveness is by this time well in evidence. Thus, of the implements already mentioned, the spear-thrower, consisting in a long bone handle with a projection at the end into which the butt of the spear fits, so that the thrower gets an extra leverage by doubling, as it were, the length of his arm, is a most ingenious device. It is not surprising that the Australian natives, as Howitt records, attribute immanent magical power, a sort of " devil," to this instrument ; and I myself have seen an Australian spear flung by this means a good 150 yards.

So, too, the harpoon heads have a swelling at the base, sometimes perforated, which shows them, if we can go by the analogy of the Eskimo weapon which is very similar in this respect, to have been detachable and connected with a line. This, held in the hunter's hand, would enable him to " play " his victim, perhaps a salmon and even more probably a seal—to judge from the fact that drawings of both these animals have come down to us from those days. The grooves, by the way, cut in some of these harpoon heads are probably not decorative but functional, and meant to hold some sort of poison.

Thanks to these same drawings we can proceed far beyond the evidence afforded by the surviving artefacts themselves in our reconstruction of a material culture which, of course, consisted largely of perishable things. Thus wooden clubs and spears of a great variety of form are depicted on the walls of the caves ; where as often as not a game animal is portrayed with one or more of such weapons sticking in some vital spot—no doubt by way of a prefiguration believed to make the prayer, or perhaps one should say the spell, come true. There are also certain mysterious pictographs which are variously interpreted as more or less symbolic representations of huts, nets, sledges and so on, and at any rate are suggestive of a many-sided mode of existence.

Meanwhile, in France and northern Spain, to which the typical Magdalenian culture is confined, it was apparently forbidden by custom, perhaps because it seemed magically dangerous, to adorn the cave walls with scenes of ordinary human life. But this taboo evidently did not apply to the contemporary inhabitants of eastern Spain—Capsians as they may be named in virtue of a presumed connexion with

North Africa. These people have left in their shelters a most vivid record of their daily avocations, and the hunting, dancing or fighting goes on before our eyes, very much as it does in the best examples of Bushman art in the South African shelters and caves.

We gather that the men were, at all events in their active moments, content to go mother-naked, if we exclude decorative trappings such as armlets, anklets and head-dresses made out of feathers. The women, on the other hand, wear seemly skirts reaching below the knee, and, though the breasts are exposed, there are signs of ornamental adjuncts to the upper part of the person ; while, for the rest, the elaborate coiffures and the attenuated waists prove the age-long appeal of Vanity Fair. Of course, when the adornments in question are of hard material, we have the originals and not merely their representations to go by. Ivory and bone, for instance, provide articles some of which were probably for purely personal decoration, such as bracelets, while for others a magico-religious significance can be assumed.

Thus, in excavating an Aurignacian site, the Paviland cave in South Wales, we came across a pendant carved from a mass of dentine that had formed at the place where a mammoth's tusk had snapped off short. Such a unique thing might well be held to be powerful " medicine " for the fortunate hunter who found it. Teeth of various kinds, bored and with incised marks on them, were worn, as we may guess from the similar practices of the modern savage, to bring about sympathetic relations, on the homoeopathic principle, with the animals of which they were once part. Skulls of the cave-bear have been carefully despoiled of their teeth in the sacred cave of Tuc d'Audoubert, and doubtless the great beast, haunter of such uncanny places, was sacred too ; though it would be more of a guess to think of it as the totem of a clan.

As for the pierced shells, which from the beginning of the Aurignacian period form a prominent feature of the grave furniture, and were not only made up into necklaces and girdles, but must evidently have been also sewn upon the garments, these too may well have served as amulets rather than as trinkets ; their presence in the graves suggesting, like the red ochre in which the bodies were laid, some mystic purpose such as the revitalising of the deceased. If it be true that some of these shells had found their way to palæolithic Europe from the Indian Ocean, some very widespread belief in the virtue of shells may be surmised.

It would perhaps be worth recalling in such a connexion that in the island-strewn seas to the east of New Guinea there is a custom of making long voyages for the purpose of exchanging, not trade goods, as a modern economist might reasonably expect, but rather luck bringers, objects made of shell such as necklaces and armlets to which a mystic virtue is attributed ; a virtue which is held to increase as the amulet acquires age.

Of all the supposed amulets of the European cave period, however, few are more plausibly identified with a sacred purpose than the objects usually described as bull-roarers. Strictly speaking, they are pendants which could never have served functionally as bull-roarers, so as to produce sounds as if of thunder or of a mighty rushing wind. Nevertheless, they may well have been imitated from the real instrument of wood, and have been regarded as symbols with equivalent powers, such as those of controlling the weather, making things grow, stimulating human fertility and so on.

Reverting to the genre paintings of the Capsians, we note a striking fact in the presence of the bow, one of the decisive triumphs of early human invention. They are fine, large bows of more than one pattern ; the arrows do not seem to be always, if ever, feathered, and the quiver is not in use. Dr. Macalister suggests that the Capsians coming up from North Africa may have been the first to introduce the bow into Europe. Certainly, the evidence for its use in Palæolithic times to the north of the Pyrenees is not clear. The lighter darts depicted as adhering to the animals' sides in the Magdalenian rock-paintings may be hand-thrown assegais ; while the Aurignacian bas-relief from Laussel which incompletely represents a male figure with left arm extended—no bow-string, however, being shown, whereas there are lines indicating a girdle—is hardly convincing, though often described as " the Archer."

So far we have been concerned with the utilitarian, though, even so, often highly decorative, side of Upper Palæolithic culture. Turning now to the remains of sculpture and painting, enough has been said to make it probable that a mystic rather than a purely æsthetic motive underlay most of these efforts, at any rate apart from eastern Spain. A difficulty to be faced, however, is that magico-religious symbolism is usually unfavourable to realism or, as it may be termed, naturalism in art—as if the letter warred with the spirit, and too great an interest in the thing itself might interfere with the higher meaning which it was intended to convey by suggestion.

Now sometimes we have every right to infer a symbolic treatment, as in the case of the female statuettes with secondary sexual characters emphasised to the point of exaggeration. Of course they may have stood for the type of beauty appreciated at the time, a sort of " Hottentot Venus " ; and there are those who detect definite steatopygy and other racial traits found among the South African Bushmen and their congeners. It is at least equally likely, however, that these figurines embody a fertility charm, and typify prolific motherhood in such a way as to bring out the idea at all costs and regardless of what actually occurs in the sensible world.

Again, in the cave of Montespan discovered in 1923 the clay figures of a bear and other animals are very rudely shaped. But, for one thing, the head of the bear is supplied by an actual skull, so that it is not improbable that the skin covered the rest ; and, for another thing, the holes with which the figure is riddled prove that stabbing formed a part of the ritual of incantation, so that no lack of realism is due to the symbolism as such.

Contrast the other cave, Tuc d'Audoubert, where clay

figures have likewise been preserved in a moist atmosphere caused by the proximity of a subterranean river. The two famous bisons, male conjoined with female, are modelled with the greatest fidelity to nature, and the accompanying rites probably involved, not a mimic slaying, but an invocation to be fruitful and multiply. The abundant prints in the neighbourhood of naked feet with heels deeply impressed show how, after the fashion of the modern savage, these early folk danced out their litanies, and doubtless could do so with greater expressiveness than they could think or speak them out.

Here, however, we are not directly concerned with the motives of primitive art, except in so far as these are reflected in, or govern, the technique. Whatever reason we assign— whether we suppose or not, for instance, that accuracy of portraiture was believed to exert a more compelling force in determining the fate of the victim—the fact remains that all the most finished work is strikingly naturalistic in tone, while the rest appears to embody less successful attempts in the same direction.

In the order of chronological development sculpture seems, as compared with painting, to have led the way, and, at any rate with the Aurignacians, to have achieved high excellence, while the sister art remained tentative and reminiscent of a clever child's first efforts. It may well be that to represent in three dimensions, that is, in the solid, comes more easily to the mind than registering a mere surface impression in two dimensions, that is, in the flat ; and one notices that a dog understands a substantive shape much better than a picture. Sculpture in relief might act as an educational bridge from the one to the other ; though this is by no means certain. Early experimenting would naturally take place at first in soft and perishable material, and so leave few traces. By a miracle the clay figures already mentioned have survived.

As for design in the flat, I have seen in the cave of Niaux a beautiful little bison drawn with a finger in the sand and preserved by a thin layer of stalagmite ; while close by was a most convincing trout executed in the same way—the ancestor, I dare say, of a delectable fish from the neighbouring stream which I had consumed that morning for breakfast. Bone, however, and in even greater degree ivory, lend themselves to carving, as do also some of the softer kinds of stone such as steatite and limestone. Out of such materials are, for instance, fashioned the figurines already mentioned, which if mostly ill-proportioned in our eyes—the head of a girl from Brassempouy and the statuette of a woman recently found at Isturitz are not so ungraceful —do not lack a certain technical merit.

Animals, however, rather than human beings would seem to have inspired the cave artist to put forth his best, and some of these studies—that, for instance, of the head of a neighing horse from Mas d'Azil—are among the world's masterpieces. The same contrast in the felicity of the rendering is noticeable in the two groups of wonderful reliefs cut in the living rock which Dr. Lalanne had the good fortune to discover. The human figures of Laussel are strongly wrought and interest us because of their possible relation to fertility cults and the like ; but they are not beautiful. At Cap Blanc close by, however, a procession of sculptured horses and bisons, eleven in all, runs round the wall of the shelter, and nothing could be finer than these reliefs, the product of rough stone tools, yet every one true to life.

To pass on from sculpture to painting, taken together with the allied art of engraving, four chief phases of development are distinguishable for the northern region—that is, for France with the Cantabrian district of Spain—only one of which is Aurignacian, the remaining three being Magdalenian. In the first phase some of the early attempts can hardly rank as art at all. Thus in the way of engraving we find so-called "arabesques" composed of a meandering series of roughly parallel lines. Those I saw on the walls and ceiling of the cave of Gargas seemed in some instances to have been simply traced with the fingers wherever the rock was faced with clay, while in others they had been scratched on the rock itself, possibly with some sort of pronged instrument. Their significance, if any, is hard to guess ; but I have elsewhere put forward the suggestion that they were intended, possibly with a totemic purpose, to imitate the actual claw-scratchings of the cave-bear, such as may be seen on the same cave walls to this day. So also, if we turn to painting, it perhaps hardly amounts to art to stencil the hand by placing it on a greased surface and blowing charcoal or red ochre from the mouth, so as to get the print in outline, as was done repeatedly in the same cave of Gargas—doubtless with some mystic purpose in view, as is further suggested by the fact that often the fingers are mutilated, as if by way of sacrifice.

True art, however, makes its first diffident appearance in the form of outline drawings of animals, either incised with a graver or in monochrome paint. Usually the body is in profile and only one front leg and one back leg are shown at first, though four legs appear before this phase is concluded. Both horns, however, are represented as if in a full-face view, since the artist has not yet learnt to distinguish between what he sees and what he merely thinks.

In the second or Lower Magdalenian phase there is altogether more life in the engraved figures. The perspective of the horns and the shaping of legs and hoofs receive due attention, while within the outline there are slight attempts at shading. Moreover, engraving now begins to be associated with painting, the monochrome line of paint being often laid fairly thickly on top of a previously engraved outline, and then cleverly broadened out so as to give the effect of shading—much in the style of a charcoal stump drawing, as Mr. Burkitt points out in his excellent account of the subject. Perhaps this is the most vigorous moment in the whole development.

In the third or Middle Magdalenian phase, engraving has become confined to rather small figures ; which in certain examples, however, such as those from Teyjat, display excellent quality. Painting has in the meantime suffered a set-back, having been seduced into experiments in flat wash and in a sort of chess-board arrangement of red and black, which were disappointing. The result is that in the fourth or Upper Magdalenian phase, whereas engraving is now confused and decadent, painting recovers something of the splendour of the second phase, but by elaborating its methods and mixing its paints ; these polychromes being likewise in some cases helped out with the graver.

Sometimes, too, there is a tendency towards impressionism, the lower parts of a mammoth at Font de Gaume appearing as a mass of shaggy hair, with a gap below across which the eye travels to four barely indicated feet. Altogether, this latest art smacks of the school rather than of nature, and may have helped to bring about its own final disappearance by becoming too self-conscious and preoccupied with style ; though doubtless there were about this time other causes, for instance climatic changes, at work to overwhelm the Magdalenian culture in general ruin.

Meanwhile this division into phases applies only to the three northern centres of artistic ascendancy, the Dordogne, the French Pyrenees and Cantabria, the last-named showing local peculiarities of a minor kind. Throughout this region the affinity of the forms of fine art is so close that I have some-times wondered whether the same artists may not have been in request over a wide area, and whether more intensive study might not enable us in some cases to recognize an individual master's hand in more than one locality.

As for eastern Spain, where there is little engraving, and the painting has both an outward style and an inner spirit of its own, a certain development here also can be traced, thanks to the custom of those days which tolerated the super-imposition of one picture on another ; as indeed did also the Bushmen, though it is said that the latter would not allow a man's work to be thus defaced for a space of three genera-tions after his death.

A word may be added about the artist's outfit. His paints, consisting of red and yellow ochre, black oxide of manganese and so on, were pounded on schist tablets which served as palettes, and were stored in hollow bones, or sometimes might be made up into a sort of pencil, looking just as if it had come from a modern stationer's shop ; a bone stylus was also used to lay on the colour. All these things have been found ; as well as the stone lamp, very like an Eskimo's, by the dim light of which an uneven face of rock, hidden away in the black interior of a mountain, was adorned with shapes as glorious as they are time-defying.

It remains to speak of those engraved or painted signs which may be supposed to embody an early attempt at writing. There can be no doubt, I think, that veritable pictographs were in use. At Niaux, for instance, in one special part of the long gallery a considerable distance away from any of the pictures of animals, one sees a large group of such marks—dots and strokes and other more complex signs distributed over the rock surface.

Amongst them one seems to make out the forms of clubs and other vague adumbrations of actual objects ; but it is clear that the general intention is to symbolise rather than to represent directly. In another part of the cave such signs occur close to the picture of a wounded bison—one, by the way, which has for its back simply a projecting gable of rock of just the right shape. Here one is tempted to read off the inscription as a spell expressive of the weapons and the encircling tactics whereby the death of the quarry is to be encompassed ; for there are marks like clubs, and circular patterns of dots which may stand for the hunters or for their tracks.

Then there are engraved marks on objects of ivory and bone, the most striking example being the ivory rod from La Crozo de Gentillo, which has been compared with an Australian message-stick. These may well have been mnemonic signs to help the bearer to remember the terms of his report, or to enable a hunter to keep tally of his victims, or perhaps in other cases to convey a spell or to indicate private property. As it is, one can but throw out bare guesses as to the more intellectual and ideal side of what spiritually, if not economically, amounted to a rudimentary civilization.

Let it be realized too that, if the caves have preserved much, still more has perished irretrievably—the wood-work, the leather-work, the feather ornaments or, again, the masks, such as the reindeer mask of the sorcerer whose portrait is to be seen in the cave of Les Trois Frères. I remember how in Honolulu, when admiring the treasures of the late dynasty, the gorgeous feather garments and the like, of which the local museum is full, I reflected that there was hardly a thing on view—the very stone-work being in a friable lava— that with normal wear and tear would leave a trace of itself in a thousand years—not to speak of the ten to fifteen thousand years that separate our time from the heyday of Upper Palæolithic culture.

The succeeding age, conveniently termed the Epipalæolithic, is from the standpoint of arts and crafts one long chapter of backwardness and squalor. The dry cold of the Magdalenian period had given way to damp and dreary conditions—a sort of protracted thaw. The cold-loving fauna had retreated north, followed possibly by the most vigorous of the hunting tribes. If any of these remained behind, they, in company with the immigrant peoples that gradually arrived, had to accommodate themselves to new material, as witness the harpoon heads of the Azilians ; these, formed out of red-deer antler, seem decidedly clumsier than those which the Magdalenians used to carve from the now unprocurable antler of the reindeer.

Fine art had perished. Probably its development was bound up with peaceful and settled conditions, such as are certainly suggested by the closely packed Magdalenian shelters of the Dordogne, where the valley of the Vézère harboured the palæolithic equivalent of a regular garden-city. On the other hand the Capsians, who herald immigration from the south, are convicted by their rock-drawings of a fondness for war, the defeated foe bristling with the arrows that have been shot into him. Promoted by these methods, invasion by the Azilian-Tardenoisian folk, who apparently follow in the footsteps of the Capsians, may have soon extirpated all taste for the gentler pursuits of life.

At most there survived, as is to be seen best in Spain, a highly stylised mode of representing men and animals, which as fine art is contemptible, but may have served some religious or pictographic purpose. The famous Azilian pebbles with their queer markings have given rise to all manner of ingenious interpretations ; but to-day the prevailing view is that these signs, some of them not unlike the letters of our alphabet, are one and all highly stylised versions of the human figure. The fact that in a cave near Basel similar pebbles had been carefully broken into fragments suggests that they may have been " soul-boxes," or life-tokens, such as modern savages sometimes store in a safe place so as to impart a corresponding safety to the individuals whom they represent.

As for stone-work, the output of this period is poor, and one must suppose that there had been a reversion to the use of wood. The little geometrically-shaped flints, angular, curved or crescentic, known as the Tardenoisian industry, and having a distribution extending from western Europe to Egypt and even India, may have had more uses than one, and only gratuitous dogmatism would lay it down for certain that they were stuck like teeth into wooden clubs, or that they served instead of fish-hooks, or that they provided a means of extracting edible snails from their shells. For the rest, though the study of this period is important as a key to the earlier movements of the present population of Europe, with its mixture of long-heads and broad-heads— the latter now appearing not only in Central Europe, but even in Portugal—the cultural remains of such midden-folk would scarcely repay a more extended notice.

In the older nomenclature the Palæolithic is immediately succeeded by the Neolithic Period, though it was always agreed that the latter began with a phase of transition which some went so far as to regard as a downright hiatus— a gap in the evidence corresponding to a state of decline and depopulation. Now that the Epipalæolithic cultures, so-called because new data, especially from Spain, prove their strict continuity with the Palæolithic, have been found

253

to cover a lengthy tract of time, it becomes a nice question how far a true Neolithic culture period is worth recognizing in itself, apart from the first-fruits of the ensuing Age of Metal.

Undoubtedly there is a considerable overlap between the two as regards Europe as a whole, sundry parts of the West and North, notably Scandinavia, using stone only, though fairly advanced in other respects, at the very time when copper and even bronze were being worked in the East and South. The fact is that the distinction between the real Stone Age of food-gatherers and the Modern Age of food-producers is not expressible in terms of the material that supplied their cutting instruments.

An economic revolution due to the domestication of animals and plants brought about a contrast in the life of Europe as great in its way as that exhibited between the habits of red man and white man in North America. The contrast was all the sharper because the new culture which was to make such short work of the old had originated outside Europe, and was at once too alien and too advanced to assimilate any essential features of the previous dispensation. Thus to explain the Robenhausian culture so-called, namely that of the Swiss lake-dwellers, with their characteristic kine, swine, cereals and so on, recourse must be had to the analogy of far-off Anau in Turkestan, where the Pumpelly Expedition found much the same things in use at a date estimated to be considerably earlier than the organization of the great river-civilizations of Mesopotamia and Egypt.

Such an hypothesis of a foreign derivation need not debar us from holding that Switzerland itself, if it did not initiate, at least fruitfully developed the new principle that Man henceforth was the master, not the slave, of Nature. When assisting at the excavation of a site on the shore of the Lake of Neuchâtel, where two settlements, both equally devoid of any trace of bronze, lay one over the other, I was much struck with the vast superiority of the later one.

Not only were the piles of the earlier lake village short and feeble, while those of the later one testified to excellent carpentry ; but from the stratification observable in the copious deposits—all the more plentiful, no doubt, because each structure in turn had been destroyed by fire—it was clear that, whereas the first group had lived strictly on local products, the very stone for their implements being poor stuff from near at hand, the second set of inhabitants had established trade relations extending hundreds of miles in at least three directions. For they had amber from the Baltic, shell necklaces from the Mediterranean, and fine homogeneous yellow flint from Grand Pressigny in France.

Indeed, when we consider this lake culture as a whole, even when abstraction is made of Bronze Age developments which lie outside the scope of the present chapter, we cannot but marvel at its richness ; a tour round the splendid museums of Switzerland being needed in order to bring the impression home. Of course, peat is a wonderful preservative, enabling us to take stock of the industry in wood ; so that we have, for instance, the hafts, wooden or else of deer-antler, in addition to the stone blade, as well as examples of their fabrics, woven materials, string-nets and basketry, and even of the grain or nuts which they ate—the last-named more or less carbonized, yet none the less possible to identify. Allowance being made for the inevitable absence of similar relics of Magdalenian times, it still remains obvious how far man has advanced in material comfort.

Even so, security of a purely economic order may be purchased dearly if as the result of increasing wealth, with corresponding density of population, there is more bitter conflict between men to offset a mitigated struggle with Nature. The lake-dweller may have been partly led to adopt his amphibious way of life by the fear of wild beasts such as wolves ; but the analogy of modern Malays or Papuans, with their pile-dwellings of very similar construction, suggests that the primary object was self-protection from human enemies. In the Swiss examples the dwellings were quite near the shore, but doubtless a removable causeway connected it with the platform on which were the huts of wattle and daub.

255

As for the frequent destruction by fire, of which there is ample evidence, this may be the work of enemies; but one must also give due weight to the fact that the primitive wooden hut, with a flat stone for its hearth, was particularly liable to such disaster. This, for instance, was also the constant fate of the old-time " long house " of the American Indian. Indeed, in general the domestic architecture of the period was flimsy and impermanent, consisting of mere cabins, often of beehive shape and partially sunk in the ground for better shelter. Here, no doubt, as in a Lapp dwelling to-day, the folk herded together, glad to be protected from the weather, and not greatly minding stuffiness and the fleas.

Incidentally the archæologist regrets the almost complete disuse of caves where a stratified deposit could collect, since in his eyes the dweller in huts, tents and other make-shift habitations stands for mere surface-finds having no bearing on the time-order.

On the other hand, in startling contrast to the meanness of the abodes of the living, the dead are housed magnificently in erections of unhewn stone that last to this day. This at least holds for the so-called Megalithic zone, which runs more or less coastwise from the Black Sea and Mediterranean round by the Atlantic and North Sea as far as the Baltic, leaving the central parts of Europe unaffected. The different types of structure comprised in the European branch of this complex—for there is another branch to the east which, if it exhibits much the same forms in Asia, diverges considerably in what are taken to be allied developments in the Pacific—are almost too well known to need detailed description.

Most typical of all is the dolmen, a single table-stone set upon uprights. The term is sometimes used to include all constructions involving a series of such cover-stones and uprights, which are otherwise known as passage-graves. The latter vary in ground-plan, the simplest being of uniform width throughout, while in other cases the passage leads into a wider chamber having, it may be, secondary cells attached.

A fine example of this more elaborate type has recently been discovered in Jersey under a tumulus known as La Hougue Bie, composed of some 20,000 tons of earth. The monument itself is roughly 75 feet long, of which about half is passage and half chamber ; the latter, with three annexes, is arched over by six capstones, each with a span of about 16 feet, and weighing up to 30 tons.

The incredible labour and architectural skill displayed in such an edifice raise questions, which science is by no means yet prepared to answer, concerning the organization and resources of the megalith builders ; the more so because in the case of the Channel Islands, which are full of fine megaliths, neither a numerous population nor opportunities of amassing wealth can well be assumed. The same holds of Malta where there are, nevertheless, structures, approximating to the megalithic type and wrought without the help of metal, which neither the Iberian Peninsula nor Brittany nor Ireland nor Scandinavia can outshine, rich as each is in such monuments.

This is not the place in which to discuss whether that preoccupation with the fate of the dead, which was so prominent a feature of Ancient Egypt, had spread, rather sooner, it would seem, than the use of copper and bronze, to the West and North by way of sea-borne trade ; or, again, how far such a cultural drift was affected by secondary developments originating in various local centres—Iberia, the Paris basin, Scandinavia, South Russia. Here we have only to take note of the existence of an architecture inspired by religious motives—the worship or at least care of the dead in the first instance, with probable evolution of the house of the dead into a temple of the gods. But in such a connexion we may take note of the rather feeble manifestations of sculpture and painting in this period.

Thus, though Malta's temples can produce sculptured forms worthy of a palæolithic artist, the Paris basin is content with rude idols representing a female with prominent breasts, presumably a fertility goddess of the type of the Great Mother of Anatolia. Painting and engraving are confined to the production of hopelessly stylized animals and human beings, which can hardly, one would think, have been intended to please the eye, though it is impossible to guess what magico-religious or pictographic meaning they had, if any.

As compared with lake-dwellings and megaliths, taken with all their cultural associations, the remaining topics that relate to the period are no doubt of minor importance. A word, however, is needed about the dawn of the art which from this point onwards affords the archæologist his best evidence for differentiating and dating the various cultures, namely, the art of ceramics.

That pottery of a rude kind is occasionally found in palæolithic sites has been asserted, and I have myself found on a Mousterian hearth portions of fire-hardened clay which a credulous person might easily take to be rudimentary potsherds. It is safer, however, to regard the potter's art as introduced into western Europe along with the rest of the food-producing tradition. Certain it is that in the Eastern Mediterranean, when copper was barely known or not at all, fine ware could be made, whereas the much later advent of bronze in the West found the pottery still coarse and wanting in variety of form. True, in connexion with the dolmens one comes upon considerable advance in ceramic refinement where it is not yet certain that metal is in use ; but in such cases it is better to give the Bronze Age the benefit of the doubt, and to suppose that one has crossed the conventional line which divides prehistory from proto-history, the Stone Age from the Bronze and Early Iron Ages.

It remains to consider the stone-work which, after all, gives the Neolithic Period its distinguishing name ; though this corresponds to a superficial distinction. To think of it as the period of polished stone is misleading for at least two good reasons ; one, that much of the stone was not polished, the other, that the process, once begun, went on well into the Age of Metal. Polishing, or at any rate grinding, which

is not quite the same thing, goes back at least to Epipalæolithic times, when it was occasionally applied to flint, after
chipping, to get a better edge.

How Palæolithic man failed to discover it I cannot think,
since Mousterian slices of greenstone show rubbed edges
which were almost certainly produced by human agency,
though no doubt accidentally in the course of some culinary
operation such as braising roots. Probably a conservative
predilection for flint prevented the earlier stone-workers
from taking advantage of a method less suitable for flint
than other kinds of stone—diabase, jadeite and so on, which
it was left to Neolithic man to appreciate. Most of these
are rather richly coloured stones which appeal to the eye,
and certainly the best Neolithic celts are meant for show at
least as much as for use. Contrast the rough Australian celts
which in most cases have no more than the edge ground,
and, however serviceable, are ugly.

In Europe, however, not only was the necessary sharpness
produced by grinding, but in the best examples the whole
surface is ground to a uniform brightness, due attention
being likewise paid to complete symmetry of shape. Probably a ceremonial use once attached to many of the choice
specimens now figuring in our collections ; and indeed in
modern Melanesia celts are set up on graves so that the
ghost may rise up and enter into them if he will.

As for the mode of manufacture, any gritty rock surface
might be used as a grindstone, and the grooves so made are
often still to be recognized. Polishing, however, especially
if the material of the celt be hard, involves far more labour
than grinding an edge, though perhaps no greater skill. Of
hafting such a celt several Neolithic fashions are known and
more may be suspected. A very simple plan, for instance,
is to bend over a flexible stick and make it fast underneath
by binding, the Australian native using gum for this purpose,
in which case notching or grooving the stone may be
employed to give the handle a better grip. But the typical
way, more especially with a blade of almond shape, is to
thrust the tapering end through a hole in a thickish wooden

haft reinforced with sinew or other binding material so as to prevent splitting.

Another method, involving perforation of the head either by pecking with a flint awl or by twirling a stick loaded with wet sand, might suit a ceremonial better than a practical use, since the stone is apt to crack at the place of boring. Neolithic drilling, by the way, can usually be detected by its " hour-glass " appearance, the work being started at both ends and stopping short of a cylindrical effect. Or, finally, the blade may simply have its butt driven into the handle, whether this be at right angles to form an axe or adze, or endwise as in a chisel.

Next in importance to the celt in a series illustrating Neolithic stonework is the arrow-head ; though, since the bow dates back to the Upper Palæolithic, while on the other hand metal, long after it was in use, was too valuable to prefix to a vagrant missile, it is far easier to determine the type than the sequence-date of a given find. It is a fair guess, however, that, when flint was first adapted to this purpose, a sharp point was of chief concern and a convenient base for attach-ment a secondary matter.

From a lozenge-shaped flake a tang which could be stuck into a hole in the shaft would evolve, by having the flanking edges chipped, until first horizontal shoulders and then recurving barbs would be formed. Such a general theory of development, however, must ignore local variations ; as when Ireland, which for reasons of its own was keenly interested in arrow-heads, exhibits certain crescentic shapes which are almost without a parallel, at any rate in western Europe.

As for other varieties of Neolithic stone-work, it would be too long a task to enumerate the forms without number which an age of increasing specialization in the arts of life was bringing into existence—an age, too, which was the heir of the previous ages and freely selected among traditional patterns. Indeed, it is often hard to distinguish from Upper Palæolithic specimens, as regards their type, implements which have a definite Neolithic horizon, or at any rate are associated with pottery and the bones of domesticated animals ; the explanation being in part, perhaps, that Capsian and post-Capsian immigration from the South brought into western Europe a tradition going back to the Aurignacian.

For the rest, there are found in England and elsewhere plenty of Neolithic implements of almost pre-Chellean coarseness, some of them doubtless unfinished instruments or rejects, while others may be for rough use in the fields as hand-picks or hoes. At the other end of the scale the Neolithic Age can compete in refinement of execution, involving pressure-flaking and the most delicate sense of symmetry, with the best Solutrean work ; so that indeed it takes the expert to distinguish the two styles. Scandinavia, in particular, where the Stone Age seems to have enjoyed a sort of St. Martin's Summer, produced masterpieces such as the Danish hand-daggers and half-moon blades which, if we leave out of account the Solutreans and the Ancient Egyptians, no other people of the ancient world could equal or even approach. Beauty apart, however, the Egyptians perhaps hold the palm for cleverness, since they could even chip flint into rings and bracelets. A famous flint-knapper of Brandon in Suffolk confessed to me that with all his modern tools he broke down at that point. Mention of Brandon, with its flint industry, which includes both mining for the material— in mines similar in all respects to ancient ones such as Grime's Graves, with its tunnelled shafts and deer-horn picks—and chipping it into the gun flints and tinder-lighters still in request in backward parts of the world, will serve, in conclusion, to remind us that even now Stone Age ways persist. They are found chiefly, of course, among savages, yet in minor matters even among ourselves ; as a visit, say, to the Hebrides would very soon show. Meanwhile, out of the Stone Age has evolved everything else.

17. ARTS AND CRAFTS OF THE MODERN SAVAGE

H aving reviewed the progress of the arts historically up to the end of the Stone Age, let us pass from archæology to ethnology, and, proceeding analytically, consider typical arts belonging to the same general level of culture as they appear among the more backward peoples of to-day. Such an analytic treatment is the only safe method to pursue in this case, since the historical antecedents of the modern savage are mostly beyond recall ; so that we cannot tell how far his present ways are genuine survivals of ancient Stone Age practices as we know them, and how far like circumstances have led him independently to converge on a like course of conduct.

Now, whereas a history takes the facts as they come, an analysis, on the other hand, must arrange its topics on a logical plan. Such a plan is suggested by the familiar contrast we are wont to draw between living and living well, work and play, necessaries and luxuries, material and spiritual interests, and so on—distinctions which all suggest that there are two planes, a lower and a higher, on which mankind lives as best it can simultaneously : the plane of the body, as it were, and the plane of the soul. Correspondingly, then, the arts of mankind can be divided into two chief classes, the Prudential and the Liberal Arts.

Some might suppose that neither Stone Age man nor his modern equivalent can have had much opportunity to develop a soul. Let them study the facts before they make up their minds on this point. Certain it is at least that in what follows we shall have as much to do with arts which in their direct purpose are not utilitarian as with arts which are ; though of course it may well be true that for Man to develop his spiritual interests for their own sake is highly practical in the long run. Thus under the general head of Prudential Arts we may range such of them as relate to food-getting and food-preparing, fighting, shelter, clothing, transport and trade ; whereas the Liberal Arts comprise those connected with what even in their earliest manifestations can be recognized as fine art, science and religion.

To begin, then, with food-getting. This consists at the lowest stage of culture in gathering, hunting and fishing, the former the woman's, the two latter the man's department. Gathering need not detain us long, since a stick with which to scratch and a receptacle of bark or skin or hollowed wood in which to store the seeds, berries, nuts and grubs constituting the find will suffice in the way of outfit. Hunting, on the other hand, in conjunction with fighting, which has much in common with it, provides so powerful a stimulus to human ingenuity—since kill or be killed is the principle involved—as almost to justify the paradox that the impulse to create is child of the impulse to destroy.

Hunting gear can be classified functionally according as its object is to strike or to catch ; while striking in its turn in all its varieties, such as battering, piercing, cutting, pelting, may be subdivided into striking at close quarters and striking at long range. To take striking weapons first, of course the hand, even an ape's undeveloped hand, can use stick or stone for beating or for throwing equally well, and up to a point the same instrument, whether club, spear or even knife, can serve either purpose indifferently. Specialization, however, soon comes into play, and, since long-range work greatly exceeds short-range in the variety of opportunity offered, invention especially triumphs in the sphere of ballistics.

Of battering appliances, the prototype is, perhaps, the stout sapling, plucked up with the root, which serves as the striking end ; such a club, slightly adapted by trimming, appears in the traditional Greek representations of Hercules. The bosses left by such rough and ready pruning of roots and side shoots are actually serviceable for offensive purposes ; and in the more developed weapons of this kind it is interesting to trace the survival of these knobs, varied in the composite type by insertions of flint spicules, shark's teeth and so on, while even decorative embellishments repeat the same theme, as in the beautiful carved maces from the Pacific.

Meanwhile, a club is capable of striking in more ways than one, and instead of bruising may have a cutting edge, resembling in this case, and perhaps deriving from, the paddle of a canoe. Or, once more, a club may pierce. I have an Australian hardwood club obtained directly from a native who showed me in vivid pantomime that the curved head was meant to strike the victim round the back of the neck, and the sharp end to finish him off by pecking him behind the ear. In this instance the fashioning has obviously been done by means of a sharp stone, and the bending, together with the hardening, of the point accomplished by the aid of fire. Of course, wood is not the only material used for clubs, though doubtless it came first. Thus, I have two weapons from Africa, one of pure ivory, the other of rhinoceros horn, and both most formidable as well as beautiful implements.

As a missile the club must be lighter and of a form lending itself to the action of throwing. The Palæolithic paintings almost certainly depict throwing-clubs of more than one shape, including that incurved shape which marks the boomerang class. Known in India, in Africa, including ancient Egypt, and even occasionally in America, as well as in Australia, where it is a leading weapon, this type has probably descended from very ancient days ; the effect of the heavier forms when sent spinning forward with an overhand action being discomforting alike to beast and to man.

Most are of the non-returning kind, and this holds even of Australia, to which the returning kind is more or less peculiar ; though Pitt-Rivers found that a model which he made of a boomerang from ancient Egypt came round and back excellently. Presumably a natural warp in the wood causing one half of the blade to deviate a little, perhaps two or three degrees, from the plane of the other directed attention to this power of orbital flight in certain implements ; whereupon selection and adaptation would step in to satisfy a taste in throwing which always, I suspect, looked chiefly to the fun of the thing, as witness the delight with which the surviving natives indulge in the practice. They love to compete with each other to see who can cast in the widest and completest circle, now from the right, now with another club from the left side ; while proud is the man whose boomerang not only completes its circle, but loops the loop, one or more times, as well—a result to which a still day, the trained hand of the thrower and, perhaps chiefly, the excellence of the weapon must all contribute.

As for the utilitarian aspect, I have heard that a returning boomerang will often take a flock of parrots unawares ; and a native told me himself that, by casting it from the bank of a stream so as to come up " all same eagle-hawk " behind a group of feeding ducks, he could make them fly low along the water into the neighbouring reeds where, immersed up to her neck, his " gin," or wife, was waiting with a net—very much as an artificial kite is sometimes trailed over partridges in England.

Next in primitiveness among striking weapons is, perhaps, the spear, originally any stick or stake with a sharp end.

According, however, to its use as a thrusting or a throwing weapon, specialization of form soon evolves. A good point, of course, is needed in any case, and, though invention supplies different materials, such as stone, bone, obsidian, a sharp end of bamboo, or, finally, metal, the shape does not greatly vary, except in so far as barbs are unsuitable for a stabbing-spear, which may have to be withdrawn quickly, bayonet-fashion. The shaft, on the contrary, is subject to considerable differentiation. To be straight and to balance well are essential in the missile spear, and much experimentation alone could bring about a conjunction of these qualities.

Even so, the perfect weapon, if thrown by hand, has a carry of little more than fifty yards. Hence the importance of the invention of the spear-thrower, or " wommera," which trebles the effective range, or, alternatively, allows a heavier spear to be used. It has already been mentioned that this ingenious instrument goes back to Magdalenian times, the rod of reindeer antler, or bone, with a hook at the end being a fairly common find. As it is often elaborately decorated with figures of animals, it may have been accredited with an inherent killing-magic. Certainly primitive man cannot fail to have been impressed and awed by the access of mechanical power produced by what Dr. Macalister calls " the first *machine* of which we have any remains invented by man."

The modern distribution of the instrument may possibly be explained by reference to a culture spread from palæolithic Europe ; but many links of the chain of evidence are missing. Of the two areas in which it occurs, one consists of Australia, with New Guinea and some neighbouring islands. A good many types co-exist in Australia, one being a simple rod and hook, which has the advantage of offering small resistance to the air, and the others being more or less blade-like. One of the latter kinds indeed has the tooth at the side of the blade so that this cuts through the air edgewise ; but the rest, with the tooth in the face, are often quite broad. Where this occurs the notion probably is that what the weapon loses as a spear-thrower it gains as a shield.

The second area is the continent of America with a slight extension into the corner of Siberia bordering on Bering Strait. The Eskimo and other " tundra " peoples have habits so like those reflected in the remains of the Magdalenians that it is tempting to postulate a historical connexion between the two ; and the common possession of a spear-thrower of very similar type is evidence in point. The examples from Central and South America bear a family resemblance to the northern type, which is characterized by special attention to the handle-grip.

Meanwhile, Europe itself abandoned the rigid type of spear-thrower in favour of a flexible one, such as the javelin-thong of the Greeks and Romans, which both increased the leverage and improved the grip. This had a loop for the fingers, was tied fast to the middle of the weapon, and flew away with it from the hand. This method survives in West Africa ; but in New Caledonia they know how to hitch on the string so that it is released and remains behind when the spear is thrown.

We pass on naturally to the subject of the bow and arrow ; for, having dealt with the spear as propelled by wood or string, we have to enquire how wood and string in combination can discharge a modified spear, an arrow being no more than a light dart which has become gradually specialized. Indeed, an arrow may be merely a sharpened stick and, as in New Guinea, lack feathers. Since such a stick as used with a plain bow of wood could leave no trace behind in a Palæolithic site, we must beware of denying the possibility of the existence of this weapon on the ground of the absence of stone or bone arrow-heads ; and indeed we should not know that the Capsians had it but for their pictures. On the other hand, Tasmania and Australia—with the exception of the extreme north of Queensland, where it intrudes from Melanesia—lack it entirely, so that it looks as if the invention belonged to a rather high level of Stone Age culture.

How this invention came about we can but guess. Perhaps, as Pitt-Rivers suggests, it was by discovering that a dart fitted to an elastic branch makes an effective spring-trap for game. When the right sort of wood is to hand, a bow of the plain or wooden variety suffices. It may, however, be simple or compound, of one kind of wood or of several kinds, as in some Japanese examples. The other main variety, known as the composite bow, made of wood, horn and sinew glued together and perhaps covered by an ornamental casing of birchbark or lacquer, is characteristic of the whole steppe-region of Asia from Turkey to China, and presumably originated in a region where wood is scarce.

The Eskimo, having to use driftwood or reindeer antler, provides the necessary elasticity by means of a backing of sinew which is simply lashed in position, a process more rudimentary and hence perhaps earlier than the Asiatic plan, found also in America, of moulding the sinew to the wood. Meanwhile, an advantage of the composite bow is that it can be made to combine great strength with moderate size, and hence lends itself to use from horseback. Thus armed the Parthian archers were the terror of the Roman soldiery; and one of Napoleon's generals recounts how he was wounded by a mounted Cossack who had nothing better than this bow to oppose to the French musket.

One might also make mention of certain unusual forms of the bow, as for instance the S-shaped plain bow peculiar to the Andamans and New Hebrides, or the reflex type of composite bow in which the ends are pulled right over into the opposite direction in stringing—no easy job. The cross-bow, however, need not detain us, since, although it is found among savages both in South-eastern Asia and in West Africa, the chances are that its invention is due to rather more advanced folk whose ideas they have borrowed.

The pellet-bow, on the other hand, is interesting because it introduces a new mode of striking—pelting instead of transfixing. The string is furnished with some sort of bag for holding the bullet, which is usually a small pebble or lump of hardened clay ; and, if he can avoid hitting his own thumb, the hunter can make short work of small game. The distribution of the instrument is perplexing, because, while found in the south-east of Asia from China as far as India, it occurs again in South America and is thus a bone of contention between those who assert and deny that Asiatic influences found their way to America across the Pacific ; denial involving the view that the pellet-bow was invented twice over.

Reverting from pellet to arrow, we may next note that a weak bow and correspondingly light arrow—a mere dart, in fact, such as the Bushmen had—will serve the hunter's purpose if he make it the vehicle of poison. The Bushmen added the venom of various snakes and caterpillars, an invention depending on direct inference. Most primitive peoples, however, use vegetable poisons, such as aconite in the Old World and curare in the New, and much experimenting with things to which magical properties were attributed, many of them of no real efficacy, must have led on to the selection of the most deadly principle. Meanwhile, any weapon—a spear or harpoon, such as the Magdalenian harpoon, perhaps, with its grooved head, or a dagger suitably perforated, or even, it is reported from some South American tribes, a thumb-nail—will convey the infection, so long as it draws blood.

Hence, from the spear through the arrow we reach the tiny poisoned dart which, wadded at its base with pith or cotton to fit the tube, is discharged from a blowpipe. With one of these weapons, nearly twice as long as himself it may be, and sometimes furnished with a sight, the Indonesian can slay the largest beasts. Such an invention, indeed, is entitled to rank with the boomerang, the spear-thrower and the bow as a first-rate achievement of primitive man.

In South-east Asia a bamboo or reed makes a natural tube, though the nodes tend to interfere with the smoothness of the barrel, despite efforts to shave them down ; so that the Bornean method of boring such a gun out of solid wood, though infinitely laborious, is perhaps found paying in the long run. In Guiana and the Amazon region, which is the other chief area in which the blowpipe is found, they are clever enough to scoop out the tube in two halves, which are then stuck together. With such an instrument they achieve very high as well as long shots in the tropical forest; and it makes no noise.

Not to deal here with cutting weapons, which have less to do with hunting than with fighting—though strictly the sharp projections and serrated edges of clubs and spears are for cutting rather than piercing—we may note that another form of pelting is by means of the sling, a very widely distributed device, though not known to Australians, Bushmen and other very primitive people. Entangling, however, rather than striking is the object of the bolas or stone balls, attached to a length of string which their weight causes to enwrap the object struck. The Eskimo have a device for catching puffins in which no less than seven or eight weights of stone or ivory are affixed to cords of sinew tied together at the other end, so that the instrument flies out like a fan, covering five or six feet. The lasso or throwing noose is an even more purely entangling weapon, and like the bolas is now mainly confined to America.

The entangling of prey in running nooses, nets and so on enters largely into the art of trapping, a subject too vast to be examined here, though the technical ingenuity involved in these and other trapping methods, as by pit, palisade, falling weight and so on, is considerable, and was doubtless exercised in Palæolithic times, even if the interpretation of certain symbols as different kinds of snare is unduly sanguine. Nor can any sketch of primitive hunting be complete without allusion to Man's success in appropriating the services of the dog—who goes back as a domestic animal to the Danish kitchen-middens, and is perhaps recognizable in the Capsian rock drawings—a hunting-companion with whom no other animal ally, cheetah, ferret, cormorant, hawk, can compare in value.

Fishing in some form is a well-nigh universal and very primitive art. At its lowest it may approximate to gathering rather than to hunting, as among the Tasmanians, who knew neither the fish-hook nor the net. The spoil of the sea took for them the form of shell-fish which the women scraped off the rocks with a wooden chisel, often diving deep for the purpose. On the other hand, as a specialized type of hunting, it develops elaborate implements and methods ; and naturally, seeing that it may become the staple industry of a whole people, such as the salmon-fishing tribes of North-West America.

Up to a point the hunter's weapons likewise serve the fisher, who similarly uses spears and arrows for striking, and nets and traps for entangling. Water tactics, however, are sufficiently different from land tactics to introduce important modifications of form—the many-pronged fish-spear, for instance, or the detachable harpoon-head, not to speak of fishing-nets and fish-traps of infinite variety. Or, again, if poisons are used to stupefy fish, they are special poisons ; or, if animal allies assist, they are special animals, the cormorant with ringed neck, the frigate-bird, the otter, the lined sucker-fish which finds the turtle ; the rarity of such experiments causing one to wonder whether man has really explored all the possibilities.

Apart from the art of diving, so well developed in tropical seas, the method most peculiar to the fishermen of all the world is the use of the line, with or without a rod, and ending in a noose, or more usually in some attachment for securing the prey, be it hook or gorge, which in order to attract must be cunningly baited, or at least be made of shining material. Prehistoric man enjoyed his salmon and trout, as Magdalenian pictures attest, but not much remains of his fishing gear, as there is little profit in trying to make a hook, still less a barbed hook, out of flint ; though I have seen one made at Brandon which caught a fish in the Suffolk Ouse. There are, however, barbed bone-implements of Magdalenian age that may be fish-hooks, and small rods of bone and ivory, sharp at both ends, which are not unlike the gorges used by the Eskimo ; these, tied in the middle, come up across the fish's mouth when the line is pulled.

271

Of particular areas with specialized fishing methods the most interesting is perhaps one extending from Indonesia to Melanesia. Here is found the fishing-kite, with its spider-web lure, which the kite causes to trip lightly along the water like a well-cast salmon-fly. Or, again, the Eskimo, who in many other ways is so well adapted to his harsh environment, has endless methods of dealing with the fish and other water animals on which he largely lives. Un-erringly he strikes the seal from his canoe with a harpoon, discharged with the spear-thrower and with bladder attached to mark the place of the kill ; or, from a larger canoe, even drives the huge whale ashore, after a dozen hunters have together fixed their harpoons in its sides.

As for the more advanced modes of food-getting involved in the pastoral and agricultural conditions, they fall rather outside the present chapter ; though of course it is true that it was the Stone Age that discovered how to get beyond the Stone Age. How the domestication of animals and plants came about we can but guess. Doubtless woman had a good deal to do with it. She probably made pets of small animals as a by-product of her mothering impulse ; and to-day among the Australians the dingo, the sole mammal except Man of long standing in that land of marsupials and mono-tremes, and almost certainly Man's companion on his first arrival, is more a pet than a useful adjunct to the hunter. Again, as gatherer-in-chief, she may have been the first to notice that, when casually dropped, the wild seeds sprouted in her untidy middens.

For the rest the pastoral life may well have developed first in Northern Asia, unless palæolithic Europe was beforehand with the taming or partial taming of the dog and the rein-deer—a process completed whenever these were broken in for sledge-transport. If traction and, later, riding were by experimentation with other animals improved into primary arts of life by male endeavour, milking would seem rather to have come from the female side as being an extension of the foster-mother idea.

So much for the results of bare conjecture, eked out by

uncertain attempts to discover the habitat of each wild species of animal and plant from which the domesticated kinds are derived.

As regards implements, the outfit of the purely pastoral tribe is meagre, being limited to what is portable, since the life is nomad. Agriculture, on the other hand, since the grower must at least stand by while the crop is in the ground, tends to be sedentary, so that gear can be accumulated. The digging stick, which perhaps as among the Bushmen has a stone ring to give weight to the head, has only to be broadened out and sharpened at the end to become a spade ; which in the metal age develops first a rim and then an entire blade of the harder substance. An even more important instrument is the hoe, at first perhaps a stone pick held in the hand, then by hafting adze-fashion converted into a hacking-tool ; though the angular stick may have developed on independent lines.

Hoe culture, which is usually woman's work, is typical of savagery. With the dawn of civilization comes plough culture, when agriculture is reinforced by the traction of pastoral animals, and the male now condescends with their aid to direct the plough, a sort of glorified hoe, along the furrow. Reaping is a difficulty before the use of metal, and is carried out with flints, usually anticipating the shape of the sickle, and with saw-like edges, which are often retained when metal comes to be used. Other processes such as threshing and winnowing also call appropriate instruments into being.

Having gathered or caught its food, the animal eats it. Not so Man, who must in most cases cook it first. Fire-making is an art that goes back at least to the earliest cave-men. Iron pyrites is found in Mousterian deposits, and, as the Fuegians obtain fire by striking together two lumps of pyrites, such a percussion-method may have been the oldest way of artificially producing fire, especially as it would be a natural by-product of a stone-chipping age. Of course Man may have used it earlier still, having obtained it from a volcano or from a tree set on fire by lightning ; and, before he knew how to create it, may have carried it about by means of a fire-brand, like Prometheus who stole fire from heaven in a reed. The Andamanese have fire, but are unable to make it.

A great benefactor of the race was the discoverer of the frictional method. The principle is always the same, namely that rubbing one piece of wood on another quickly enough will start a spark in the sawdust. Perhaps it was found out accidentally in the course of trying to bore a hole through a soft bit of wood with a harder stick. Pushing along a groove is one frictional method, characteristic of Oceania. Sawing, as when an Australian scrapes his spear-thrower across his shield, or when, from Malaya to Melanesia, a flexible strip of bamboo is used, is another quite effective way. But the commonest plan is to twirl the spindle-stick with the hands. I have seen an Australian produce fire in this way in under half a minute ; though with the very appliances that he had used I could myself by no means approach his record.

Increased rotation is produced by the thong-drill of the Eskimo, or by their bow-drill, which is an improvement on the thong-drill, since it can be worked with one hand instead of two. Finally, in Indonesia, they discovered, perhaps in the course of making blowpipes out of bamboo, that by driving a piston home into a bamboo tube a spark will be generated in tinder placed at the bottom. This compressed-air method concludes the list of primitive ways of making fire, unless one includes the use of a concave mirror, whereby the ancient Mexicans obtained fire from heaven on ceremonial occasions.

Methods of cooking provide a large subject which must be merely glanced at in passing. Despite the popular belief to the contrary, the savage much prefers his food cooked to raw, and even his cruder methods, such as baking the meat encased in clay over hot stones, the whole well covered over with earth—a plan often adopted by our gypsies—yield savoury results ; while, when the pit-oven has been evolved, delicacies both animal and vegetable are produced which the white man has often been glad to copy.

Hot water is more of a difficulty, owing to the want of a receptacle that will stand the fire. For this purpose the most primitive folk use calabashes, wooden bowls, skins, water-tight baskets and so on, and the " stone-boiling " method of throwing in hot stones has to serve. As for basketry, which in its two main forms is either woven, or coiled and then sewn—though there are numerous sub-varieties, such as those that may be studied in the admirable work character-istic of western North America from Alaska to California—it has the advantage of providing more or less nomad peoples with vessels that both are light and will not break. Though a really well-made basket will hold water, it is advisable to reinforce it with an exterior coating of clay ; and it may have been by discovering that such a containing shell on exposure to heat became fireproof that the idea of pottery arose.

This is indeed one of the ways in which the primitive potter proceeds, namely, to build up his clay round some object, a wickerwork frame or a gourd ; though the drawback is that the object in question suffers when the pot is fired. A commoner and better method is to hollow out a lump of clay, shaping its inside with a stone and its outside with a mallet ; or else to make a disk for the base and carry up the walls by adding strips, which when long are carried round in a spiral coil. These spirals are easily smoothed or beaten out when the vessel is complete, though they may be retained for the sake of ornament, as found in some examples of Pueblo ware.

The decoration of pottery, however, is too vast and intricate a subject to enter into here. Nevertheless it may be noted that some primitive band-patterns are but reminiscences of the marks left by a rope or a plaited band such as may be tied round the pot to support it during the baking. For the rest, pottery up to the invention of the potter's wheel is, like basketry, mostly woman's work, and this should be remembered to her credit when we contemplate the beautiful developments of ceramics, not indeed found among the rudest peoples, but associated with those who have just left savagery behind and are at that intermediate stage of culture which is termed barbarism.

To pass on to fighting, this activity on its offensive side draws largely on the same types of weapon as does hunting : club, throwing-stick, spear, arrow being but slightly modified for purposes of war. The axe, which otherwise is more of a carpentering than a hunting implement, develops into the battle-axe chiefly after metal has come into use. Metallurgy hardly comes within the scope of this chapter, but in the present context a word may be said about the development of cutting instruments, such as tend to encourage the hand-to-hand organized combat—the most deadly form of fighting, at any rate in early times, and one to which very primitive folk are not prone.

A piece of raw copper, hammered cold, as by the Eskimo, is no great improvement on the stone celt which it slavishly copies. When bronze is invented—or one might perhaps say discovered, since tin occurs with copper in natural alloys which do not, however, provide that ten per cent. of tin which is the optimum admixture—casting in moulds either of clay or of stone, single or if necessary double, enables a hard-edged weapon such at the palstave to be constructed ; together with suitable fittings, flange or socket, for firm attachment to the haft, which was always a weak point with the stone axe.

Special to war, again, is the whole large class of defensive appliances. Lacking a tough skin, Man has been forward to utilize the hides of other animals in the way both of shield and cuirass. Curiously enough the value of the turtle's carapace as a buckler is a hint from Nature of which he does not seem to have taken much advantage. The light parrying-sticks of Australia testify to active as contrasted with passive self-defence, an art which the natives have cultivated to perfection.

The helmet develops from the head-dress, which originally

is meant to frighten at least as much as to protect. Thus the
traditional lion-skin of Hercules proclaimed him equipped
with the lion's virtue ; and indeed a great part of the
warrior's outfit is magico-religious rather than utilitarian
in its prime intention. This, however, as well as many other
aspects of the art of war, must be passed by ; as, for instance,
the whole subject of the trophies of war, including those
gruesome spoils of the head-hunter which fill our ethnological
museums.

Shelter, the next of the primary needs of Man which the
prudential arts subserve, should be considered in close
relation to war, as the defensive value of a site for habitation
must have influenced selection ever since the right to a snug
cave was disputed with the cave-bear. The history of
fortification has quite an intricate first chapter, since even
primitive warfare necessitates the choice of special situa-
tions, hill-tops, swamps and so on, and the use of special
devices such as stockades, earthworks, ditches and pile-
dwellings, whether on land or in the water.

As for shelter from the elements, the Tasmanian wind-screen
or Australian " wurley " is hardly more adequate a dwelling
than a hare's form. Nomad conditions, however, especially
if unassisted by animal transport, preclude anything more
substantial than such a tent as can be put together with
the help of a few poles and some skins, mats or pieces of
bark. Thus the Eskimo when hunting during the summer
uses a flimsy lodge of skins stretched over whalebone. In
the winter, however, he settles down in his warm if stuffy
"igloo" built of blocks of snow laid in spiral courses; or else
digs himself a pit-dwelling six feet below ground, and
domed a little above ground-level with earth supported on
whale-ribs or driftwood.

Farther south the Indian dwelling, whether the portable
tent or " tipi " of the plains or the fixed hut or " wigwam "
of the more sedentary tribes, was a humble affair if intended
for a single family. The " long house," on the other hand,
which sheltered a number of families, as among the Iroquois,

might be one hundred feet in length, and was stoutly constructed of poles covered with brush matting. Meanwhile, the Pueblo or Village Indians, using the sun-dried brick known as adobe, built structures of many cells and storeys, the lowest storey being a more or less blank wall which one scaled by ladders that could be drawn up in times of danger. The elaborate stone buildings of Mexico and the other regions of high culture in America are perhaps no more than developments of this hive-like type of architecture. These examples, taken from one continent, will suffice to illustrate the variety of the domestic forms of habitation.

To house the spirits of the dead, or the gods, with due regard for their comfort and dignity is an idea that may well have grown out of the practice, dating from Mousterian times, of burying the dead man in the floor of his own cave-dwelling. In the end, however, it comes to entail a host of special constructions outside the scope of this brief survey.

Clothing may seem at first sight to be an elementary need of Man, but it is possible that self-decoration, with attractiveness to the other sex as a conjoined motive, is the mainspring of the effort to improve on Nature's garb. We have only to think of the Capsian rock-paintings, where the men sport magnificent head-dresses, armlets, anklets and so on, but are otherwise stark naked ; while the women, if more fully attired, can scarcely be supposed to have taken to skirts because they felt the climate more.

Even conditions of extreme cold do not make warm garments a necessity for our species ; since, whereas the Eskimo of the Arctic rejoices in heavy furs, the Fuegian braves the Antarctic blasts with no other protection for his bare person than a skin slung to windward as a screen. And, as climate will not account for the degree in which a primitive people is clothed, so neither will the general state of their culture, Africa, for instance, being far more addicted to nudity than North America, though comparing favourably with it in the matter of social development.

As regards the leading processes whereby clothes, as con-

trasted with ornaments, are made, we may put aside
feather-work, bead-work, quill-work and so on as ministering
to the latter interest, and attend solely to two types of
industry, namely leather-work and weaving. To take leather
first, skins of animals are in almost universal use as a
covering against the weather or as armour ; while tents,
boats, harness, boxes, cradles, shields, in fact a good half
of the entire gear of the savage, will usually be found to
involve this material, to which every beast of reasonable
size contributes.

Skin dressing is a rather elaborate business. The multitude
of implements of the scraper class, with a more or less
specialized form, in which a prehistoric site abounds can
be referred, by analogy with the ways of the modern savage,
to the different requirements of each stage in an operation
which includes stripping, scraping, rubbing, pounding,
squeezing, drying, dressing in various ways as by tanning,
greasing or smoking, and finally softening, which may be
assisted by unlimited chewing, as among the Eskimo women.
For sewing leather the bone needle is ineffective, but the
awl can produce quite neat results. For footwear, apart
from occasional experiments such as the grass sandal,
leather is in general demand, and from the mere bag of
hide the sole sewed to an upper of softer skin has developed ;
though, on the whole, it is remarkable how many peoples
of considerable culture retain the barefoot habit.

Next, as regards weaving, it may be noted that in the
broadest sense this term covers the production of all textiles,
including basketry, mats and wickerwork. We may, on
the other hand, confine it to meaning the work of the loom.
The first step towards this mechanism is to hang the warp
on the branch of a tree, from which it is an easy step to set
up poles supporting a warp-beam. Sometimes the weaver's
fingers suffice to introduce the weft strands, as in the case
of the beautifully-patterned Chilkat blankets from Alaska ;
but various artificial aids—the shuttle, originally a mere
stick round which the thread is wound so as to pass it
through the warp, the weaver's sword, or else a comb to

drive back the weft threads, and so on—are gradually devised.

To pass to the subject of transport, this too is a primary need of Man, especially when he is more or less of a nomad and must carry about with him the whole of his earthly possessions. It is mostly the woman's job at this early stage to load herself up with the household chattels, including the baby, as the man has to keep his hands free to use his weapons against wild beast or human foe.

We seem to discern sledges in some prehistoric designs, but cannot tell whether these were hand-drawn affairs, or whether the idea of harnessing some sort of drag to dog or reindeer had yet occurred to the hunter. The toboggan lying flat on the ground is hardly more rudimentary in conception than a couple of trailing sticks with cross-bars, out of which, more especially when ice-work was contemplated, the sledge with runners would develop, a contrivance which among the Eskimo becomes a miracle of neat and effective workmanship.

The travois of the Prairie Indians, on the other hand, first used with the dog, but with the horse as soon as the latter was introduced by the Europeans, was but a two-pole drag, being sometimes hardly more than the rolled-up " tipi " or skin-tent tied on the top of two bunches of tent-poles. Apart from freighting purposes, however, such conveyances are soon found useful for the transport of the aged, sick, wounded, children and puppies ; so that riding presently comes into view as a distinct ambition for art to oblige.

The wheel, peculiar to the Old World, probably develops out of the wooden roller used for the transport of stones, as by the megalith-builders. It would be noted that the centre could be conveniently pared down till the axle in one piece with its solid wheels, as still to be seen in Asia Minor, would emerge. Then at last, becoming lighter, the wheel would be made to turn on a fixed axle.

Passing from land to water, we do not know who it was that first tried to cross a stream on a log, but the use of such a float, probably at first not so much a vehicle as an aid to swimming, would easily develop into the raft, which is no more than a compound float—at any rate until the mere platform, at water level and consequently uncomfortably wet, is supplemented with some erection designed to surround the passengers with a sort of water-tight screen.

To a large extent the form of the early boat is governed by the material available. The Eskimo has to depend on skin for his "kayak," or for the larger "umiak," the women's boat. Farther south, where the birch grows, the Indian uses the bark-canoe, hardly suitable for more than river-work. The dug-out, again, pre-supposes a stout log, and even where Nature provides the largest trees an upper limit is reached which forbids development on the scale attained by the plank-boat. The latter may have developed out of the custom of adding a gunwale to the dug-out to increase its size and make it more seaworthy, until the accessory technique took charge of the construction from top to bottom. Dug-outs, by the way, go back at least to Neolithic times. The outrigger canoe, which extends from India to the Pacific, may have originated in the joining of two dug-outs together.

The subject of trade connects naturally with that of transport, since to be able to exchange goods implies a power of moving a certain superfluity of stock. At this point, however, we are perhaps across the border-line that divides strict need from the pursuit of enjoyment, as the evidence seems to show that prehistoric man was at least as prone to treat himself to imported luxuries, such as shells or amber for his person, as, say, to a more workable kind of flint for his tools.

The history of exchange has an important aspect which cannot be dealt with here, since it relates to the spiritual side of the transaction, namely, to the establishment of friendly relations between different groups of normally hostile men. What is known as the silent trade, where each party deposits something that he is prepared to barter in a neutral spot, and the tacit haggling is concluded when each carries off what the other has offered, illustrates the difficulty of coming to an understanding with the stranger.

For trading purposes a great step in advance is made when such understanding takes the form of a recognition of a standard value, to which more uncertain values can be referred, so that a medium of exchange or currency comes into use. Thus any raw material in universal demand, such as tobacco, tea or salt, and especially metal such as copper or iron, provides a currency when made into sticks, bricks or bars representing a more or less definite amount. Even in primitive Australia red ochre seems to have a sort of currency value. The domesticated animal, again, among all pastoral peoples, affords a ready way of calculating the price of a marriage or of a murder, as is shown by the very derivation of the word pecuniary from the Latin " pecus," cattle.

All these, however, are useful objects, whether exchanged in their natural state or in some manufactured form, as when metal in the shape of hoe-blades, knives or armlets is so used. Of ornamental objects, shells have the greatest vogue as currency, though there are other possibilities in this direction, as Melanesia proves, where not only strings of shell-beads, but whale's teeth, feather-work and so on are also in circulation. Most famous of all shell-currency is the cowrie, which, though native to certain small islands in the Indian Ocean, counts as money in Asia as well as right across Africa.

For the rest, a recent study by Dr. Malinowski of the so-called trading-voyages of the Melanesians shows that economic reasons as we understand them do not inspire them so much as motives, very hard for us to follow, concerned with the honour and the luck that attend the giving and receiving of objects of ceremonial value, such as certain necklaces and armlets of shell. It looks as if even at the primitive level Man is aware that the good things of life include more than the supports and comforts of physical existence. To this other aspect, then, of the arts as aiders and abettors of his nascent spiritual nature let us now turn.

That the liberal arts, the humanities, the pursuits that embody Man's aspiration towards beauty and truth and goodness, in a word, towards the ideal, should be utterly separate and distinct from those prudential activities which aim directly at self-maintenance is unthinkable, in view of the fact that the best way to succeed in life is to feel it to be worth living. It might seem, indeed, that the savage is too hard put to it in his struggle with his untamed environment to have time or inclination for occupations in which the interest in fact is subordinated to a free indulgence in fancy. Yet those who know him best maintain that, so far from resembling the practical man of the modern world, he is typically a mystic—one who projects his emotions into things so that they stand in his eyes rather for what they symbolize than for what they are.

The sentimental values, so to speak, prevail over the sensible values. An Australian native living in the central deserts seems to us, who judge by his miserable surroundings, the most unfortunate of human beings ; yet, when the curtain is lifted to disclose the drama of his inner life, we find that every bare stick and stone is charged with traditional associations that make him appear to himself to participate in mysteries of infinite significance and hence to be most blest.

In every ethnological collection there is a vast array of artefacts vaguely labelled "ceremonial objects" on the ground that they have no obvious use, and must be, so to speak, stage properties, appropriate to some play of which the text is lost. In short, Man is and always has been body and soul, realist and idealist, at once and together.

Beauty, as we have seen, came with the dawn. Men, however, it would appear, are not all equally endowed with the power of appreciating it ; for Neolithic man is to Upper Palæolithic very much as Roman to Greek. Meanwhile, those who have this divine sense seem capable of framing a vision of the absolutely beautiful, even if it be limited to some single aspect of it. The cave artists were hunters whose keen eyes perceived the grace of form and movement in the wild creatures about them, and, if they slew, they likewise admired.

True, we have good reason to suspect that the caves were sanctuaries, and that the paintings were part of a ritual designed to secure good hunting. Even so, the artists wrought too faithfully to be merely rated as the makers of spells ; and we have the rest of their material culture to bear witness to the passion with which they sought to endow everything that they owned with the charm of perfect form.

But enough has been said already about prehistoric art, whether plastic or graphic, and it only remains to add that among modern savages there is but one close parallel, namely the art of the Bushmen of South Africa, which may quite possibly be the lineal descendant of the other. The caves of South Africa, as compared with those of France, are shallower and more like shelters, so that the Bushman had the advantage of more light for his work ; yet, on the whole, he was not so good a painter as Magdalenian man. But there is one curious development, of which traces occur in prehistoric art, which he had carried to perfection, namely a pecking method whereby the rock surface is made to represent shading as well as outline.

There is reason to think, though the evidence is poor, that the Bushman was trying to cast a spell on the animals he was depicting. When human beings are shown, they seem generally to be engaged in ritual dances and, in particular, to be impersonating the beasts that they hunted—those which they knew well being, by the way, much more accurately rendered than the cattle introduced to their notice by later immigrants, native or European. For the rest, we must be content here to take stock of a few of the more striking examples of plastic skill exhibited by primitive folk.

West African wood-carving, for instance, has lately attracted much attention. More remarkable still is the bronze-work from Benin. Even if we suppose that the " cire-perdue " method was introduced by the Portuguese—a process whereby the wax model, after being cast in a clay mould with a duct, is melted out and the metal introduced in its place—the cleverness of the native designs is not to be gainsaid. Or, again, the stone statues of Easter Island, some of them thirty feet high, rank among the world's marvels. How and by whom they were fashioned remains something of a mystery, but there are Melanesian analogies which would perhaps account for the odd cast of human countenance favoured—one that seemingly aims at representing Man with some of the attributes of the sacred frigate-bird. Meanwhile, throughout the Pacific wood is carved with the most exquisite taste, as notably in the way of house decoration among the Maori and other Polynesians.

We must, however, pass by manifestations of the fine arts which have already received some notice in other contexts to consider a hitherto unmentioned type, namely the art of music. Durable materials do not enter into the composition of musical instruments, so that nothing survives of this kind from prehistoric times, with the doubtful exception of certain pierced phalanges of reindeer which, used as whistles, give out much the same sound as can be extracted from a key.

On the other hand, all mankind is given to singing, partly to help on the work in hand in the fashion of a sailors' chanty and partly to relieve the feelings. Whether such feelings be joyful or the reverse, rhythm by regulating the expression of emotion causes self-mastery, and hence brings comfort ; wherefore the need for rhythm is primary, whereas melody develops later. A sing-song chant punctuated by drumming with the fists on opossum rugs stretched tightly across the knees forms the musical accompaniment to which an Australian corrobboree is danced with immense spirit. Indeed, drumming on a skin or piece of wood, together with knocking two sticks together in the manner of a castanet, would seem to be as elementary a way of producing rhythmical sounds as any, so that percussion instruments are probably at least as early as wind instruments, while the remaining group of string instruments may well be later than the other two. Apart from certain classes of percussion instruments, such as rattles and jingling contrivances that do not lend themselves to much elaboration, there are important varieties such as the drum, or, again, the bell, the interest of which is not confined to music, since they are useful for many other purposes as well, notably for signalling. A signalling drum, by the way, made by hollowing out an immense tree, and looking like a large canoe, is worth a journey to the Naga Hills of Assam to behold. The friction drum, characteristic of Africa, is another curious development.

Melody, as distinguished from rhythm, is most successfully attained by the xylophone, which in the best African specimens is capable of a graded series of the most delicate notes. It is to be noted how a resonator in the shape of a gourd or other hollow body has been added—a clear example of invention. Vibrating instruments of the type of the jew's-harp can be made of wood, especially of bamboo, and are widely distributed among primitive peoples.

Wind implements might originate in very simple experiments with holed bone or cane or shell. The flute and the trumpet in all their diverse forms eventually emerge. As an ingenious invention for producing a more continuous stream of wind than the cheek can supply, the bagpipe deserves notice. Its wide distribution, not only covering Europe, but extending to India and again to North Africa, suggests that it is of ancient origin ; and, indeed, its appearance in Scotland can hardly be dissociated from the earlier wave of Celtic immigration.

String instruments might be suggested by the twanging of any taut sinew, but undoubtedly the bow has had much to do with this particular development. It may be noted that the Bushmen, who likewise used the drum and a peculiar form of reed pipe, had a four-stringed harp evolved out of the bow, as well as a kind of dulcimer based on a combination of twelve bows. Their intense love of music shows that one kind of artistic excellence does not interfere with another, but, on the contrary, may well join with it in fruitful alliance.

To rest the bow on a hollow vessel, or, better, to fix such a resonator to the instrument, starts the musical bow on its career. The harp is differentiated at the outset simply by using a number of small bows with one large resonator such as a gourd.

Before leaving the subject of the cult of beauty, we must deal with a manifestation of it which may not have the same paramount interest for us as for primitive folk, but must none the less be recognized as one of the chief factors in Man's spiritual progress. Self-adornment, however, is a little hard to disentangle from the clothing of the body for protection's sake. Again, it is often difficult to say how much is to be ascribed to æsthetic motives and how much to religious, as when amulets, masks, vestments and so on are worn to bring luck, or the better to maintain a sacred character or position.

In the latter case, however, the distinction of aspects matters less because in one way or another a spiritual rather than a utilitarian object is sought. The fact is that primitive morality has hardly got beyond the ideal of cutting a good figure in society. It is not enough that a man should mean well ; he must appear well in the eyes of his neighbours. A good man is known by his stately pose, his impressive manner, his open pride in himself. Hence to such a type of the noble savage as the North American brave his decorations are precious, forming, as it were, no small part of his moral make-up. His very religion favours the same insistence on outward display, since, so far as his ornaments are luck bringers, they proclaim him in touch with the occult powers, a help to friends and a terror to enemies.

To exclude the utilitarian element the more completely, let us consider a branch of the art of self-adornment into which this element hardly enters ; one, too, that is thoroughly primitive, inasmuch as it can flourish only where nakedness prevails. This is the art of denaturalizing the body, to use a term wide enough to cover its many varieties. Into some of these we need not go at length ; such as the artificial deformation of the skull, either by flattening, or by lateral pressure, causing elongation ; or, again, the mutilation of lip, nose, ear, tooth, hand, foot and other organs, mostly for the purpose of attaching decorative adjuncts such as labrets, nose-plugs, and so on, but sometimes for religious reasons. But much might be said about the well-nigh universal practice of tattooing, taken together with the cicatrisation or scarring, which mostly serves as an alternative with black-skinned peoples.

The custom goes back to prehistoric times, to judge by the marks on the engraved woman's figure of Solutrean date from Predmost in Moravia ; while female figurines of the earliest Egyptian dynasties are similarly adorned. Presumably men painted their bodies before it dawned on them that by rubbing the paint over a previously punctured skin they could render the effect permanent. Possibly, too, since red ochre from prehistoric times onwards has been associated with burials, being no doubt symbolic of the blood needed to revivify the dead, there was all along a religious side to the tattooer's art, which was considered capable of imparting through infusion of colour an enhanced vitality to the living.

A recent work on the history of tattooing, by Mr. Hambly, tries to bring out the many-sided efficacy of a mystic kind imputed to the practice by primitive folk. As for beauty, one cannot refuse to see a certain grace in the symmetrical lines and spirals of the Maori " moko " ; while in Japan the art, though confined to the lower classes of society, has developed a technique which surpasses all other known styles in delicacy of form and richness of colour.

Passing from the pursuit of beauty to that of truth, it might seem that science and philosophy are late products of human development. Though this is by no means true, it is not perhaps easy to illustrate the beginnings of the quest for knowledge from the side of the arts. For such information we must look chiefly to the oral traditions of savages—to the myths dealing with the creation of the world, with the origin of man and his institutions, in short with everything about which an intelligent child with his wondering " why ? " wants to know.

Nevertheless, science rests on a twofold foundation-stone which was laid down in primitive times, namely, on the art of recording facts coupled with the art of measuring them exactly. First, then, as to early attempts to make and keep records. The rude marks and symbols which prehistoric man painted or carved on stone and bone, as well as doubtless on more perishable materials like wood, bark, the hide of animals and his own skin, have little significance for us, and indeed, if they could be interpreted, would possibly be found to refer mostly to passing events of slight importance ; yet in them lies the germ of writing and of the all-important substitution of a recorded for an oral tradition of culture.

Such signs are all mnemonic in the sense that they serve as aids to memory ; but it is convenient to reserve this term for those which are more or less without suggestive quality, such as knots in a string like the " quipu " of Peru, or the notches on the wooden tallies once common, and still occasionally used, in England. Such a distinction, however, must not be applied too rigidly, since the " wampum," or shell-beads of North America, though primarily mnemonic, could be made by varying the arrangement or the colour to suggest vaguely both particular objects and even abstract ideas such as peace and war.

On the other hand, the pictograph which, either realistically, or conventionally by emphasizing outstanding traits, presents a given thing to the eye belongs to the far more interesting group of symbolic signs, from which our alphabet is ultimately derived. From pictograph to ideograph is but a short step, and indeed any combination of pictographs implies that some sort of general idea is being expressed. The decisive step in advance occurs when a phonetic system, a syllabary or an alphabet, comes into being. Thereupon the syllable or sound is represented by a sign that no doubt always originally stood for a concrete object with a name embodying the syllable or sound in question. Often, as in Egyptian or Chinese, the scribe trusted his phonetic symbol so little that he added a determinative picture of the thing as it looked to help out the word as it sounded.

To turn to the art of measuring, a great deal might be said about its development by the aid of a mechanism which man always has had at his disposal, namely, his body. Thus, for purposes of numeration, he had simply to count by his fingers in tens, or by fingers and toes in scores ; and it is a common mistake to suppose that, because separate words are lacking in a primitive language for numbers higher than, say, three, therefore the power of reckoning in larger sums was absent.

Ciphering by means of objects such as seeds or pebbles— the very word calculate coming from the Latin " calculus," a pebble—is no hard matter to invent ; though arranging them on a counting-board, especially if this be furnished with a blank or " zero " column, involves considerable ingenuity. Again, spatially, the finger, hand, foot, the natural yard from the middle of the chest to the middle of the outstretched hand, and the natural fathom from hand to hand across the chest, are measures easy to apply.

Time-reckoning, on the other hand, is perhaps not very important to the early hunter—far less at any rate than it is to the agriculturist. For the hunter it is time to eat when you have killed, and time to sleep when you have eaten. Seasons no doubt affect him in a general way, and he realizes, especially in northern latitudes, that the year goes round ; while moons, since moonlit nights are an advantage, are more closely watched, and form the basis of the hunter's calendar.

As for the time of day, every hunter can fix it well enough for his purposes by reference to the habits of animals and plants, as we ourselves also judge by cock-crow or by flowers that open and shut. A surer method, however, is to watch the waxing and waning of the shadows ; and it is but a step from the natural sun-dial to the post or pillar set up in an open place with marks round it to show where the shadow will fall at the different hours. Another form of clock is provided by anything that can be reckoned on to complete a process in a given time, as a candle or lamp that burns itself out, or a vessel that fills or empties itself with water or sand.

When an agricultural year coinciding with the return of the seasons has to be worked out, both sun and stars must be observed carefully, and astronomical lore becomes an important part of religion. Thus there can be little doubt that the cultivators of the Bronze Age paid great attention to orientation, and by marking the direction with stones could predict where sun or star would appear at stated times ; being probably under the impression that by means of their rites they were not so much recording as actively regulating the movement of the universe.

Moreover, not only the physical sciences, but likewise that other important group, the biological, can be traced back to very primitive beginnings, as is best seen by studying the origins of the allied arts of medicine and surgery. Something has already been said about poisons. It remains to add that simples of beneficent virtue are no less in request, and that of a list of these collected from Australian natives a very large number were found actually to have the remedial properties claimed.

As for surgery, the savage is a good patient, and it is wonderful what cures such rough instruments can effect ; as is illustrated by the successful cases of trepanning which are often reported from the primitive world—from New Britain, for instance, where crowns cracked by sling-stones are constantly mended in this way. Nay, every hunter possesses what Tylor calls a good " butcher's anatomy," and can

operate on the body of his wounded mate, with due regard
for the positions and functions of the various parts.

It remains to say something, from the standpoint of the arts,
about the furtherance of the higher life or primitive man as
effected through his religion.   On a superficial view his
rites and the accompanying beliefs might seem to us childish,
when they are not positively degrading.   But, since all
religious feeling expresses itself through a certain symbolism,
we must be sure before passing harsh judgments on other
men's forms of worship that we know what they mean for
them.   A stick or a stone, a piece of bread or a smear of
blood—something in itself trifling or even vile—may stand
for ineffable mysteries.

The case of the bull-roarer is instructive.   Here is a mere
slat of wood which when whirled round at the end of a string
produces a queer booming sound.   In itself it seems worthy
only to be a toy, and such it is to-day in most parts of
England, though its Scots name of " thunderspell " suggests
that it has not yet quite parted with its mystic associations.
It is usually held to go back to Palæolithic times, since certain
ivory pendants of Magdalenian age look very like miniature
bull-roarers that have become amulets.   Be this as it may,
the distribution of the instrument is so world-wide that with
some reason it has been hailed as the most far-travelled, and
therefore most ancient, religious symbol in the world.
Perhaps a utilitarian value helped to gain it mystic significance ;   for animals are surprisingly scared by the sound,
so that the Bushman hunter uses it to round up his game,
the Malay planter to drive away elephants from his crops,
and the European peasant to call his cattle home, or rather
to send them flying towards the byre.

Meanwhile, the sound of the bull-roarer might well suggest
the noises of wind and rushing rain and thunder, very
welcome noises in a dry land.   So it is no wonder if the
Australian hears in it the voice of a great being who causes
all things to grow.   Such a High God, who is likewise in all
other things the friend of man and has given him all his

institutions, presides throughout South-eastern Australia over the initiation ceremonies at which the boys are turned into men—very appropriately, since he is the god who makes all things to grow. Thus round a thing of naught, a small boy's buzzer as we rank it in England now, has developed a religion of remarkable intensity and elevation among a people whose material culture is that of the Stone Age.

It would be out of the question here to attempt the briefest survey of primitive religious symbols in all their variety. The savage is so much of a mystic that religion is bound up with almost every act of his life. In fact, all that has been said hitherto about his arts should be taken as subject to the qualification that, whatever obviously practical ends they may serve, they have also uses, largely hidden from us, which relate wholly to the world of the spiritual and occult.

A spear, for instance, may be fitted with a piece of human bone which, apart from any mechanical value that it may possess, ensures the assistance of a friendly ghost who can make the weapon fly true. Indeed, one might overhaul a savage warrior from top to toe without finding a single article of his equipment that was not intended so much to frighten and bedevil his enemy as to wound him ; or, conversely, as much to sustain the soul of the wearer as to protect his body.

Or let him suppose a chief to be building himself a fine house of carved wood. He is not thinking out his plans in terms of mere accommodation or display. On the contrary, from the moment of driving in the first post—which he will be careful to establish by a foundation-sacrifice involving the pouring of blood into the hole, perhaps of human blood —he is concerned largely with the construction of devices calculated to bring a blessing or, what comes to the same thing, to avert the evil eye. Many a hideous image which is put down to the discredit of the idolater is merely a sort of lightning-conductor—something that by its very grotesqueness compels the ill-natured passer-by to bestow his first glance upon it and so, as it were, to let off his piece at a dummy.

Apart from the endless cases of a magico-religious purpose associated and almost indistinguishably mixed with practical and worldly ends, there is a whole vast group of what are usually classed as " ceremonial " objects. It is just as well to employ some such neutral term, since to decide what is religious and what merely magical in the material accompaniments of a savage rite is hard, and indeed depends entirely on one's private definition of these debatable terms. Certain it is that, by simple inspection of the thing in question, and without full knowledge of the motive, whether spiritually uplifting or degrading, that underlies its use, one could not tell under which category to place it.

For example, the museums of Australia abound with strange devices—fantastically shaped arrangements of sticks and grass and feathers—which are associated with various ceremonies ; and these are certainly religious in the sense that the natives declare that their performance not only brings them food and other material advantages but likewise makes the participators feel " strong " and " glad " and " good." In the case of Central Australia we happen to have very sound and detailed evidence about the meaning of such rites, which fall into two main divisions : those intended to cause Nature to be fruitful; and those commemorative of ancestral beings of the Golden Age who are conceived, almost after Darwin's manner, as half-human, half-animal, though in any case infinitely glorious, and such as it is soul-enlarging to imitate and portray in solemn dance.

Or, again, masks would deserve most careful treatment, not only because of their prehistoric origin and world-wide distribution, but perhaps especially in view of the fact that the masked human figure almost certainly develops into the cult image. Thus, apart from the tendency, already noticed, of the artist to weave two designs into one, there is historical reason for that confusion of human and animal forms which runs through so much of the religious art of the savage. The famous sorcerer, for instance, of the cave of Les Trois Frères has his upper parts concealed by a huge and impressive mask representing the reindeer ; but the legs nevertheless bewray the man.

As fine art such hybridizations of forms which Nature keeps apart are doubtless unpleasing. As religion, however, they are perfectly suitable for those who extract from a complexity of symbols, as we do from a group of diverse letters, an idea of many, yet harmonious, aspects.

So much, then, by way of final reminder that, in order to appreciate savage arts and crafts, it is not enough to stare at a row of museum cases. Some previous attempt must have been made to understand primitive life as a whole. And this, after all, is not so hard as it may seem. For these, too, are human beings, nay, are of the very fashion of our own ancestors. In short, if in their case different conditions of life have provoked different reactions, nevertheless Man at heart is ever much the same. Hence, for the sympathetic and fair-minded, the way to a just appreciation of the long and successful effort of our race to achieve self-culture and self-realization lies always open.

INDEX

297